Introduction to Classical Mechanics

Introduction to Classical Mechanics

Alexis Callahan

Larsen & Keller
www.larsen-keller.com

Introduction to Classical Mechanics
Alexis Callahan
ISBN: 978-1-64172-684-9 (Hardback)

Larsen & Keller
Published by Larsen and Keller Education,
5 Penn Plaza,
19th Floor,
New York, NY 10001, USA

Cataloging-in-Publication Data

Introduction to classical mechanics / Alexis Callahan.
p. cm.
Includes bibliographical references and index.
ISBN 978-1-64172-684-9
1. Mechanics. 2. Physics. I. Callahan, Alexis.
QC125.2 .I58 2022
531--dc23

For more information regarding Larsen and Keller Education and its products, please visit the publisher's website www.larsen-keller.com

Table of Contents

Preface

It is with great pleasure that I present this book. It has been carefully written after numerous discussions with my peers and other practitioners of the field. I would like to take this opportunity to thank my family and friends who have been extremely supporting at every step in my life.

Classical mechanics deals with the motion of macroscopic objects such as projectiles, parts of machinery as well as the astronomical objects including planets, stars, spacecrafts, etc. Newton's three laws of motion form the basis of classical mechanics. Some of the other forces studied within this field are Lorentz force for electromagnetism and gravitational force. It is broadly classified into three main branches such as statics, dynamics and kinematics. It can also be divided on the basis of the region of its application such as celestial mechanics, relativistic mechanics, statistical mechanics and continuum mechanics. Classical mechanics assumes that matter and energy have definite attributes, such as location, in speed and space. This book provides comprehensive insights into the field of classical mechanics. Some of the diverse topics covered herein address the varied branches that fall under this category. The extensive content of this book provides the readers with a thorough understanding of the subject.

The chapters below are organized to facilitate a comprehensive understanding of the subject:

Chapter - What is Classical Mechanics?

Classical mechanics is the discipline of physics that deals with the motion of projectiles, spacecrafts, planets, stars, galaxies, etc. It is based on Newton's laws of mechanics. This is an introductory chapter which will introduce briefly all the significant aspects of classical mechanics.

Chapter - Kinematics

The branch of classical mechanics which determines the motion of objects, points, and systems of bodies without considering the external forces is known as kinematics. Distance, displacement, speed, velocity, acceleration and motion are studied under it. This chapter has been carefully written to provide an easy understanding of these aspects related to kinematics.

Chapter - Force and Newton's Laws of Motion

The action that changes the motion of an object is known as force. Conservative force, centripetal force , centrifugal force, inertia and newton's laws of motion are few of the concepts that fall within its domain. The topics elaborated in this chapter will help in gaining a better perspective about these types of force as well as Newton's laws of motion.

Chapter - Rotational Motion

The motion of a rigid body that occurs in a circular orbit about an axis with a common angular velocity is known as rotational motion. Some of its aspects are moment of inertia, angular momentum, torque, rolling, etc. This chapter has been carefully written to provide an easy understanding of these various aspects related to rotational motion.

Chapter - Work and Energy

In physics, the work is defined as the product of force and displacement. The capacity of doing work is termed as energy. It includes potential, kinetic, thermal, electrical, chemical and nuclear energy. This chapter closely examines the concepts of work and energy to provide an extensive understanding of the subject.

Chapter - Gravitation

The universal force of attraction which acts between all matter with mass and energy including planets, stars, galaxies, light, etc. is referred to as gravitation. It includes gravitational constant, gravitational acceleration, gravitational potential, etc. All these diverse concepts of gravitation have been carefully analyzed in this chapter.

Alexis Callahan

1

What is Classical Mechanics?

Classical mechanics is the discipline of physics that deals with the motion of projectiles, spacecrafts, planets, stars, galaxies, etc. It is based on Newton's laws of mechanics. This is an introductory chapter which will introduce briefly all the significant aspects of classical mechanics.

Classical mechanics is used for describing the motion of macroscopic objects, from projectiles to parts of machinery, as well as astronomical objects, such as spacecraft, planets, stars, and galaxies. It produces very accurate results within these domains, and is one of the oldest and largest subjects in science, engineering and technology.

Besides this, many related specialties exist, dealing with gases, liquids, and solids, and so on. Classical mechanics is enhanced by special relativity for objects moving with high velocity, approaching the speed of light; general relativity is employed to handle gravitation at a deeper level; and quantum mechanics handles the wave-particle duality of atoms and molecules.

In physics, classical mechanics is one of the two major sub-fields of study in the science of mechanics, which is concerned with the set of physical laws governing and mathematically describing the motions of bodies and aggregates of bodies. The other sub-field is quantum mechanics.

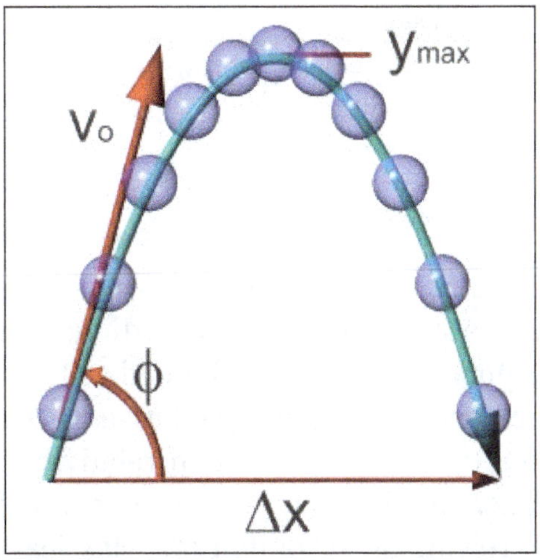

The analysis of projectile motion is a part of classical mechanics.

The following introduces the basic concepts of classical mechanics. For simplicity, it often models real-world objects as point particles, objects with negligible size. The motion of a point particle is characterized by a small number of parameters: its position, mass, and the forces applied to it. Each of these parameters is discussed in turn.

In reality, the kind of objects which classical mechanics can describe always have a non-zero size. (The physics of very small particles, such as the electron, is more accurately described by quantum mechanics). Objects with non-zero size have more complicated behavior than hypothetical point particles, because of the additional degrees of freedom—for example, a baseball can spin while it is moving. However, the results for point particles can be used to study such objects by treating them as composite objects, made up of a large number of interacting point particles. The center of mass of a composite object behaves like a point particle.

Displacement and its Derivatives

The SI derived units with kg, m and s	
Displacement	m
Speed	m s^{-1}
Acceleration	m s^{-2}
Jerk	m s^{-3}
Specific energy	m^2 s^{-2}
Absorbed dose rate	m^2 s^{-3}
Moment of inertia	kg m^2
Momentum	kg m s^{-1}
Angular momentum	kg m^2 s^{-1}
Force	kg m s^{-2}
Torque	kg m^2 s^{-2}
Energy	kg m^2 s^{-2}
Power	kg m^2 s^{-3}
Pressure	kg m^{-1} s^{-2}
Surface tension	kg s^{-2}
Irradiance	kg s^{-3}
Kinematic viscosity	m^2 s^{-1}
Dynamic viscosity	kg m^{-1} s

The displacement, or position, of a point particle is defined with respect to an arbitrary fixed reference point, O, in space, usually accompanied by a coordinate system, with the reference point located at the origin of the coordinate system. It is defined as the vector r from O to the particle. In general, the point particle need not be stationary relative to O, so r is a function of t, the time elapsed since an arbitrary initial time. In pre-Einstein relativity (known as Galilean relativity), time is considered an absolute, i.e., the time interval between any given pair of events is the same for all observers. In addition to relying on absolute time, classical mechanics assumes Euclidean geometry for the structure of space.

Velocity and Speed

The velocity, or the rate of change of position with time, is defined as the derivative of the position with respect to time or,

$$\vec{v} = \frac{d\vec{r}}{dt}.$$

In classical mechanics, velocities are directly additive and subtractive. For example, if one car traveling East at 60 km/h passes another car traveling East at 50 km/h, then from the perspective of the slower car, the faster car is traveling east at 60 – 50 = 10 km/h. Whereas, from the perspective of the faster car, the slower car is moving 10 km/h to the West. Velocities are directly additive as vector quantities; they must be dealt with using vector analysis.

Mathematically, if the velocity of the first object is denoted by the vector $\vec{u} = u\vec{d}$ and the velocity of the second object by the vector $\vec{v} = v\vec{e}$ where u is the speed of the first object, v is the speed of the second object, and and $\vec{e}$ are unit vectors in the directions of motion of each particle respectively, then the velocity of the first object as seen by the second object is:

$$\vec{u}' = \vec{u} - \vec{v}$$

Similarly:

$$\vec{v}' = \vec{v} - \vec{u}$$

When both objects are moving in the same direction, this equation can be simplified to:

$$\vec{u}' = (u - v)\vec{d}$$

Or, by ignoring direction, the difference can be given in terms of speed only:

$$u' = u - v$$

Acceleration

The *acceleration*, or rate of change of velocity, is the derivative of the velocity with respect to time (the second derivative of the position with respect to time) or,

$$\vec{a} = \frac{d\vec{v}}{dt}.$$

Acceleration can arise from a change with time of the magnitude of the velocity or of the direction of the velocity or both. If only the magnitude, υ, of the velocity decreases, this is sometimes referred to as *deceleration*, but generally any change in the velocity with time, including deceleration, is simply referred to as acceleration.

Frames of Reference

While the position and velocity and acceleration of a particle can be referred to any observer in any state of motion, classical mechanics assumes the existence of a special family of reference frames in terms of which the mechanical laws of nature take a comparatively simple form. These special reference frames are called inertial frames. They are characterized by the absence of acceleration of the observer and the requirement that all forces entering the observer's physical laws originate in identifiable sources (charges, gravitational bodies, and so forth). A non-inertial reference frame is one accelerating with respect to an inertial one, and in such a non-inertial frame a particle is subject to acceleration by fictitious forces that enter the equations of motion solely as a result of its accelerated motion, and do not originate in identifiable sources. These fictitious forces are in addition to the real forces recognized in an inertial frame. A key concept of inertial frames is the method for identifying them. For practical purposes, reference frames that are unaccelerated with respect to the distant stars are regarded as good approximations to inertial frames.

The following consequences can be derived about the perspective of an event in two inertial reference frames, S and S', where S' is traveling at a relative velocity of $\vec{u}$ to S:

- $\vec{v}' = \vec{v} - \vec{u}$ (the velocity $\vec{v}'$ of a particle from the perspective of S' is slower by $\vec{u}$ than its velocity $\vec{v}$ from the perspective of S).
- $\vec{a}' = \vec{a}$ (the acceleration of a particle remains the same regardless of reference frame).
- $\vec{F}' = \vec{F}$ (the force on a particle remains the same regardless of reference frame).
- The speed of light is not a constant in classical mechanics, nor does the special position given to the speed of light in relativistic mechanics have a counterpart in classical mechanics.
- The form of Maxwell's equations is not preserved across such inertial reference frames. However, in Einstein's theory of special relativity, the assumed constancy (invariance) of the vacuum speed of light alters the relationships between inertial reference frames so as to render Maxwell's equations invariant.

Forces

Newton was the first to mathematically express the relationship between force and momentum. Some physicists interpret Newton's second law of motion as a definition of force and mass, while others consider it to be a fundamental postulate, a law of nature. Either interpretation has the same mathematical consequences, historically known as "Newton's Second Law":

$$\vec{F} = \frac{\mathrm{d}\vec{p}}{\mathrm{d}t} = \frac{\mathrm{d}(m\vec{v})}{\mathrm{d}t}.$$

The quantity $m\vec{v}$ is called the (canonical) momentum. The net force on a particle is, thus, equal to rate change of momentum of the particle with time. Since the definition of acceleration is $\vec{a} = \frac{\mathrm{d}\vec{v}}{\mathrm{d}t}$, when the mass of the object is fixed, for example, when the mass variation with velocity found in special relativity is negligible (an implicit approximation in Newtonian mechanics), Newton's law can be written in the simplified and more familiar form,

$$\vec{F} = m\vec{a}.$$

So long as the force acting on a particle is known, Newton's second law is sufficient to describe the motion of a particle. Once independent relations for each force acting on a particle are available, they can be substituted into Newton's second law to obtain an ordinary differential equation, which is called the *equation of motion*.

As an example, assume that friction is the only force acting on the particle, and that it may be modeled as a function of the velocity of the particle, for example:

$$\vec{F}_{\mathrm{R}} = -\lambda\vec{v}$$

with λ a positive constant. Then the equation of motion is,

$$-\lambda\vec{v} = m\vec{a} = m\frac{\mathrm{d}\vec{v}}{\mathrm{d}t}.$$

This can be integrated to obtain,

$$\vec{v} = \vec{v}_0 e^{-\lambda t/m}$$

where $\vec{v}_0$ is the initial velocity. This means that the velocity of this particle decays exponentially to zero as time progresses. In this case, an equivalent viewpoint is that the kinetic energy of the particle is absorbed by friction (which converts it to heat energy in accordance with the conservation of energy), slowing it down. This expression can be further integrated to obtain the position $\vec{r}$ of the particle as a function of time.

Important forces include the gravitational force and the Lorentz force for electromagnetism. In addition, Newton's third law can sometimes be used to deduce the forces acting on a particle: if it is known that particle A exerts a force $\vec{F}$ on another particle B, it follows that B must exert an equal and opposite *reaction force*, -$\vec{F}$, on A. The strong form of Newton's third law requires that $\vec{F}$ and -$\vec{F}$ act along the line connecting A and B, while the weak form does not. Illustrations of the weak form of Newton's third law are often found for magnetic forces.

Energy

If a force $\vec{F}$ is applied to a particle that achieves a displacement $\Delta\vec{s}$, the *work done* by the force is defined as the scalar product of force and displacement vectors:

$$W = \vec{F}\cdot\Delta\vec{s}.$$

If the mass of the particle is constant, and W_{total} is the total work done on the particle, obtained by summing the work done by each applied force, from Newton's second law:

$$W_{\text{total}} = \Delta E_k,$$

where E_k is called the kinetic energy. For a point particle, it is mathematically defined as the amount of work done to accelerate the particle from zero velocity to the given velocity v:

$$E_k = \frac{1}{2}mv^2.$$

For extended objects composed of many particles, the kinetic energy of the composite body is the sum of the kinetic energies of the particles.

A particular class of forces, known as *conservative forces*, can be expressed as the gradient of a scalar function, known as the potential energy and denoted E_p:

$$\vec{F} = -\vec{\nabla} E_p.$$

If all the forces acting on a particle are conservative, and E_p is the total potential energy (which is defined as a work of involved forces to rearrange mutual positions of bodies), obtained by summing the potential energies corresponding to each force,

$$\vec{F}\cdot\Delta\vec{s} = -\vec{\nabla} E_p\cdot\Delta\vec{s} = -\Delta E_p \Rightarrow -\Delta E_p = \Delta E_k \Rightarrow \Delta(E_k + E_p) = 0.$$

This result is known as *conservation of energy* and states that the total energy,

$$\sum E = E_k + E_p$$

is constant in time. It is often useful, because many commonly encountered forces are conservative.

Beyond Newton's Laws

Classical mechanics also includes descriptions of the complex motions of extended non-pointlike objects. The concepts of angular momentum rely on the same calculus used to describe one-dimensional motion.

There are two important alternative formulations of classical mechanics: Lagrangian mechanics and Hamiltonian mechanics. These, and other modern formulations, usually bypass the concept of "force," instead referring to other physical quantities, such as energy, for describing mechanical systems.

Classical Transformations

Consider two reference frames S and S'. For observers in each of the reference frames an event has space-time coordinates of (x, y, z, t) in frame S and (x', y', z', t') in frame S'. Assuming time is measured the same in all reference frames, and if we require $x = x'$ when $t = 0$, then the relation between the space-time coordinates of the same event observed from the reference frames S' and S, which are moving at a relative velocity of u in the x direction is:

$$x' = x - ut$$

$$y' = y$$

$$z' = z$$

$$t' = t$$

This set of formulas defines a group transformation known as the Galilean transformation (informally, the Galilean transform). This group is a limiting case of the Poincaré group used in special relativity. The limiting case applies when the velocity u is very small compared to c, the speed of light.

For some problems, it is convenient to use rotating coordinates (reference frames). Thereby one can either keep a mapping to a convenient inertial frame, or introduce additionally a fictitious centrifugal force and Coriolis force.

Limits of Validity

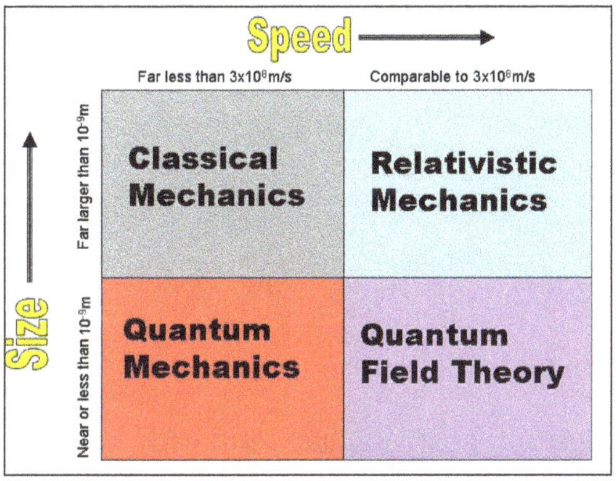

Domain of validity for Classical Mechanics.

Many branches of classical mechanics are simplifications or approximations of more accurate forms; two of the most accurate being general relativity and relativistic statistical mechanics. Geometric optics is an approximation to the quantum theory of light, and does not have a superior "classical" form.

The Newtonian Approximation to Special Relativity

Newtonian, or non-relativistic classical momentum,

$$\vec{p} = m_0\vec{v}$$

is the result of the first order Taylor approximation of the relativistic expression:

$$\vec{p} = \frac{m_0\vec{v}}{\sqrt{1-v^2/c^2}} = m_0\vec{v}\left(1+\frac{1}{2}\frac{v^2}{c^2}+\dots\right), \text{ where } v=|\vec{v}|$$

when expanded about,

$$\frac{v}{c}=0$$

so it is only valid when the velocity is much less than the speed of light. Quantitatively speaking, the approximation is good so long as,

$$\left(\frac{v}{c}\right)^2 << 1$$

For example, the relativistic cyclotron frequency of a cyclotron, gyrotron, or high voltage magnetron is given by,

$$f = f_c\frac{m_0}{m_0+T/c^2},$$

where f_c is the classical frequency of an electron (or other charged particle) with kinetic energy T and (rest) mass m_0 circling in a magnetic field. The (rest) mass of an electron is 511 keV. So the frequency correction is 1 percent for a magnetic vacuum tube with a 5.11 kV. direct current accelerating voltage.

The Classical Approximation to Quantum Mechanics

The ray approximation of classical mechanics breaks down when the de Broglie wavelength is not much smaller than other dimensions of the system. For non-relativistic particles, this wavelength is,

$$\lambda = \frac{h}{p}$$

where h is Planck's constant and p is the momentum.

Again, this happens with electrons before it happens with heavier particles. For example, the electrons used by Clinton Davisson and Lester Germer in 1927, accelerated by 54 volts, had a wave length of 0.167 nm, which was long enough to exhibit a single diffraction side lobe when reflecting from the face of a nickel crystal with atomic spacing of 0.215 nm. With a larger vacuum chamber, it would seem relatively easy to increase the angular resolution from around a radian to a milliradian and see quantum diffraction from the periodic patterns of integrated circuit computer memory.

More practical examples of the failure of classical mechanics on an engineering scale are conduction by quantum tunneling in tunnel diodes and very narrow transistor gates in integrated circuits.

Classical mechanics is the same extreme high frequency approximation as geometric optics. It is more often accurate because it describes particles and bodies with rest mass. These have more momentum and therefore shorter De Broglie wavelengths than massless particles, such as light, with the same kinetic energies.

2
Kinematics

The branch of classical mechanics which determines the motion of objects, points, and systems of bodies without considering the external forces is known as kinematics. Distance, displacement, speed, velocity, acceleration and motion are studied under it. This chapter has been carefully written to provide an easy understanding of these aspects related to kinematics.

Kinematics is a branch of physics and a subdivision of classical mechanics concerned with the geometrically possible motion of a body or system of bodies without consideration of the forces involved (*i.e.,* causes and effects of the motions).

Kinematics aims to provide a description of the spatial position of bodies or systems of material particles, the rate at which the particles are moving (velocity), and the rate at which their velocity is changing (acceleration). When the causative forces are disregarded, motion descriptions are possible only for particles having constrained motion—*i.e.,* moving on determinate paths. In unconstrained, or free, motion, the forces determine the shape of the path.

For a particle moving on a straight path, a list of positions and corresponding times would constitute a suitable scheme for describing the motion of the particle. A continuous description would require a mathematical formula expressing position in terms of time.

When a particle moves on a curved path, a description of its position becomes more complicated and requires two or three dimensions. In such cases continuous descriptions in the form of a single graph or mathematical formula are not feasible. The position of a particle moving on a circle, for example, can be described by a rotating radius of the circle, like the spoke of a wheel with one end fixed at the centre of the circle and the other end attached to the particle. The rotating radius is known as a position vector for the particle, and, if the angle between it and a fixed radius is known as a function of time, the magnitude of the velocity and acceleration of the particle can be calculated. Velocity and acceleration, however, have direction as well as magnitude; velocity is always tangent to the path, while acceleration has two components, one tangent to the path and the other perpendicular to the tangent.

Distance

Distance is a scalar quantity, which means the distance of any object does not depend on the direction of its motion. The distance of an object can be dened as the complete path travelled by an object. E.g.: if a car travels east for 5 km and takes a turn to travel north for another 8 km, the total distance travelled by car shall be 13 km. The distance can never be zero or negative and it is always more than the displacement of the object. The distance of the object gives complete complete information about the path travelled by the object.

Let's understand with the following diagram,

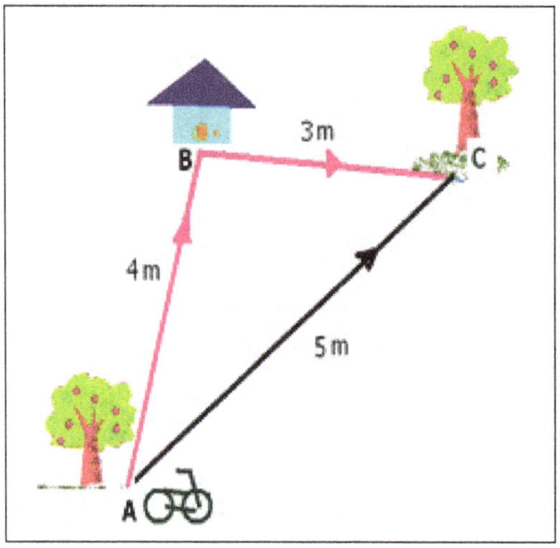

Distance here will be = 4m +3m +5m = 12 m.

Distance formula,

$$\Delta d = d_1 + d_2$$

Displacement

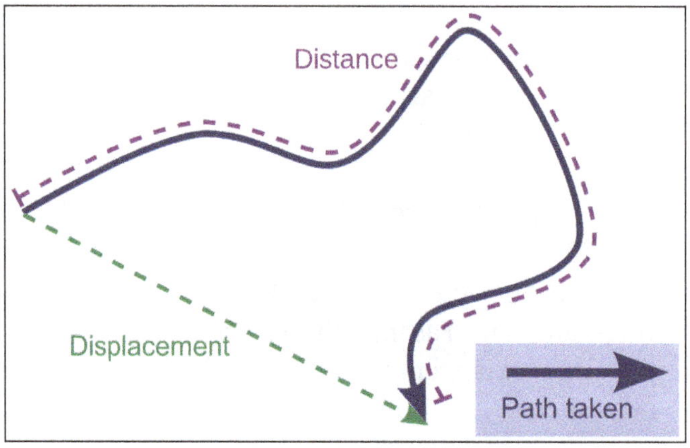

Displacement versus distance traveled along a path.

A displacement is a vector whose length is the shortest distance from the initial to the final position of a point P. It quantifies both the distance and direction of an imaginary motion along a straight line from the initial position to the final position of the point. A displacement may be identified with the translation that maps the initial position to the final position.

A displacement may be also described as a 'relative position', that is, as the final position x_f of a point relatively to its initial position x_i. The corresponding displacement vector can be defined as the difference between the final and initial positions:

$$s = x_f - x_i = \Delta x$$

In considering motions of objects over time, the instantaneous velocity of the object is the rate of change of the displacement as a function of time. The instantaneous speed, then, is distinct from velocity, or the time rate of change of the distance traveled along a specific path. The velocity may be equivalently defined as the time rate of change of the position vector. If one considers a moving initial position, or equivalently a moving origin (e.g. an initial position or origin which is fixed to a train wagon, which in turn moves with respect to its rail track), the velocity of P (e.g. a point representing the position of a passenger walking on the train) may be referred to as a relative velocity, as opposed to an absolute velocity, which is computed with respect to a point which is considered to be 'fixed in space' (such as, for instance, a point fixed on the floor of the train station).

For motion over a given interval of time, the displacement divided by the length of the time interval defines the average velocity. (Note that the average velocity, as a vector, differs from the average speed that is the ratio of the path length—a scalar—and the time interval.)

Rigid Body

In dealing with the motion of a rigid body, the term displacement may also include the rotations of the body. In this case, the displacement of a particle of the body is called linear displacement (displacement along a line), while the rotation of the body is called angular displacement.

Derivatives

For a position vector **s** that is a function of time t, the derivatives can be computed with respect to t. The first two derivatives are frequently encountered in physics.

Velocity:

$$\mathbf{v} = \frac{d\mathbf{s}}{dt}$$

Acceleration:

$$\mathbf{a} = \frac{d\mathbf{v}}{dt} = \frac{d^2\mathbf{s}}{dt^2}$$

Jerk:

$$\mathbf{j} = \frac{d\mathbf{a}}{dt} = \frac{d^2\mathbf{v}}{dt^2} = \frac{d^3\mathbf{s}}{dt^3}$$

These common names correspond to terminology used in basic kinematics. By extension, the higher order derivatives can be computed in a similar fashion. Study of these higher order derivatives can improve approximations of the original displacement function. Such higher-order terms are required in order to accurately represent the displacement function as a sum of an infinite series, enabling several analytical techniques in engineering and physics. The fourth order derivative is called jounce.

Speed

In everyday use and in kinematics, the speed of an object is the magnitude of its velocity (the rate of change of its position); it is thus a scalar quantity. The average speed of an object in an interval of time is the distance travelled by the object divided by the duration of the interval; the instantaneous speed is the limit of the average speed as the duration of the time interval approaches zero.

Speed has the dimensions of distance divided by time. The SI unit of speed is the metre per second, but the most common unit of speed in everyday usage is the kilometre per hour or, in the US and the UK, miles per hour. For air and marine travel the knot is commonly used.

The fastest possible speed at which energy or information can travel, according to special relativity, is the speed of light in a vacuum $c = 299792458$ metres per second (approximately 1079000000 km/h or 671000000 mph). Matter cannot quite reach the speed of light, as this would require an infinite amount of energy. In relativity physics, the concept of rapidity replaces the classical idea of speed.

Italian physicist Galileo Galilei is usually credited with being the first to measure speed by considering the distance covered and the time it takes. Galileo defined speed as the distance covered per unit of time. In equation form, that is

$$v = \frac{d}{t}$$

where v is speed, d is distance, and t is time. A cyclist who covers 30 metres in a time of 2 seconds, for example, has a speed of 15 metres per second. Objects in motion often have variations in speed (a car might travel along a street at 50 km/h, slow to 0 km/h, and then reach 30 km/h).

Instantaneous Speed

Speed at some instant, or assumed constant during a very short period of time, is called *instantaneous speed*. By looking at a speedometer, one can read the instantaneous speed of a car at any instant. A car travelling at 50 km/h generally goes for less than one hour at a constant speed, but if it did go at that speed for a full hour, it would travel 50 km. If the vehicle continued at that speed for half an hour, it would cover half that distance (25 km). If it continued for only one minute, it would cover about 833 m.

In mathematical terms, the instantaneous speed v is defined as the magnitude of the instantaneous velocity **v**, that is, the derivative of the position r with respect to time:

$$v = |\mathbf{v}| = |\dot{r}| = \left|\frac{d\mathbf{r}}{dt}\right|$$

If s is the length of the path (also known as the distance) travelled until time t, the speed equals the time derivative of s:

$$v = \frac{ds}{dt}$$

In the special case where the velocity is constant (that is, constant speed in a straight line), this can be simplified to $v = s / t$. The average speed over a finite time interval is the total distance travelled divided by the time duration.

Average Speed

Different from instantaneous speed, *average speed* is defined as the total distance covered divided by the time interval. For example, if a distance of 80 kilometres is driven in 1 hour, the average speed is 80 kilometres per hour. Likewise, if 320 kilometres are travelled in 4 hours, the average speed is also 80 kilometres per hour. When a distance in kilometres (km) is divided by a time in hours (h), the result is in kilometres per hour (km/h).

Average speed does not describe the speed variations that may have taken place during shorter time intervals (as it is the entire distance covered divided by the total time of travel), and so average speed is often quite different from a value of instantaneous speed. If the average speed and the time of travel are known, the distance travelled can be calculated by rearranging the definition to,

$$d = \bar{v}t$$

Using this equation for an average speed of 80 kilometres per hour on a 4-hour trip, the distance covered is found to be 320 kilometres.

Expressed in graphical language, the slope of a tangent line at any point of a distance-time graph is the instantaneous speed at this point, while the slope of a chord line of the same graph is the average speed during the time interval covered by the chord. Average speed of an object is Vav = s÷t.

Difference between Speed and Velocity

Speed denotes only how fast an object is moving, whereas velocity describes both how fast and in which direction the object is moving. If a car is said to travel at 60 km/h, its speed has been specified. However, if the car is said to move at 60 km/h to the north, its velocity has now been specified.

The big difference can be discerned when considering movement around a circle. When something moves in a circular path and returns to its starting point, its average velocity is zero, but its average speed is found by dividing the circumference of the circle by the time taken to move around the circle. This is because the average velocity is calculated by considering only the displacement between the starting and end points, whereas the average speed considers only the total distance travelled.

Tangential Speed

Linear speed is the distance travelled per unit of time, while tangential speed (or tangential velocity) is the linear speed of something moving along a circular path. A point on the outside edge of a merry-go-round or turntable travels a greater distance in one complete rotation than a point nearer the center. Travelling a greater distance in the same time means a greater speed, and so linear speed is greater on the outer edge of a rotating object than it is closer to the axis. This speed along a circular path is known as tangential speed because the direction of motion is tangent to the circumference of the circle. For circular motion, the terms linear speed and tangential speed are used interchangeably, and both use units of m/s, km/h, and others.

Rotational speed (or angular speed) involves the number of revolutions per unit of time. All parts of a rigid merry-go-round or turntable turn about the axis of rotation in the same amount of time. Thus, all parts share the same rate of rotation, or the same number of rotations or revolutions per unit of time. It is common to express rotational rates in revolutions per minute (RPM) or in terms of the number of "radians" turned in a unit of time. There are little more than 6 radians in a full rotation (2π radians exactly). When a direction is assigned to rotational speed, it is known as rotational velocity or angular velocity. Rotational velocity is a vector whose magnitude is the rotational speed.

Tangential speed and rotational speed are related: the greater the RPMs, the larger the speed in metres per second. Tangential speed is directly proportional to rotational speed at any fixed distance from the axis of rotation. However, tangential speed, unlike rotational speed, depends on radial distance (the distance from the axis). For a platform rotating with a fixed rotational speed, the tangential speed in the centre is zero. Towards the edge of the platform the tangential speed increases proportional to the distance from the axis. In equation form:

$$v \propto r\omega,$$

where v is tangential speed and ω is rotational speed. One moves faster if the rate of rotation increases (a larger value for ω), and one also moves faster if movement farther from the axis occurs (a larger value for r). Move twice as far from the rotational axis at the centre and you move twice as fast. Move out three times as far and you have three times as much tangential speed. In any kind of rotating system, tangential speed depends on how far you are from the axis of rotation.

When proper units are used for tangential speed v, rotational speed ω, and radial distance r, the direct proportion of v to both r and ω becomes the exact equation,

$$v = r\omega.$$

Thus, tangential speed will be directly proportional to r when all parts of a system simultaneously have the same ω, as for a wheel, disk, or rigid wand.

Units

Units of speed include:

- Metres per second (symbol m s^{-1} or m/s), the SI derived unit;
- Kilometres per hour (symbol km/h);
- Miles per hour (symbol mi/h or mph);
- Knots (nautical miles per hour, symbol kn or kt);

- Feet per second (symbol fps or ft/s);
- Mach number (dimensionless), speed divided by the speed of sound;
- In natural units (dimensionless), speed divided by the speed of light in vacuum (symbol *c* = 299792458 m/s).

Conversions between common units of speed					
	m/s	km/h	mph	knot	ft/s
1 m/s =	1	3.6	2.236936	1.943844	3.280840
1 km/h =	0.277778	1	0.621371	0.539957	0.911344
1 mph =	0.44704	1.609344	1	0.868976	1.466667
1 knot =	0.514444	1.852	1.150779	1	1.687810
1 ft/s =	0.3048	1.09728	0.681818	0.592484	1

Examples of Different Speeds

Speed	m/s	ft/s	km/h	mph	Notes
Approximate rate of continental drift	0.00000001	0.00000003	0.00000004	0.00000002	4 cm/year. Varies depending on location.
Speed of a common snail	0.001	0.003	0.004	0.002	1 millimetre per second
A brisk walk	1.7	5.5	6.1	3.8	
A typical road cyclist	4.4	14.4	16	10	Varies widely by person, terrain, bicycle, effort, weather
A fast martial arts kick	7.7	25.2	27.7	17.2	Fastest kick recorded at 130 milliseconds from floor to target at 1 meter distance. Average velocity speed across kick duration
Sprint runners	12.2	40	43.92	27	Usain Bolt's 100 metres world record.
Approximate average speed of road cyclists	12.5	41.0	45	28	On flat terrain, will vary
Typical suburban speed limit in most of the world	13.8	45.3	50	30	
Taipei 101 observatory elevator	16.7	54.8	60.6	37.6	1010 m/min
Typical rural speed limit	24.6	80.66	88.5	56	
British National Speed Limit (single carriageway)	26.8	88	96.56	60	
Category 1 hurricane	33	108	119	74	Minimum sustained speed over 1 minute
Speed limit on a French autoroute	36.1	118	130	81	

Highest recorded human-powered speed	37.02	121.5	133.2	82.8	Sam Whittingham in a recumbent bicycle
Muzzle velocity of a paintball marker	90	295	320	200	
Cruising speed of a Boeing 747-8 passenger jet	255	836	917	570	Mach 0.85 at 35000 ft (10668 m) altitude
The official land speed record	341.1	1119.1	1227.98	763	
The speed of sound in dry air at sea-level pressure and 20 °C	343	1125	1235	768	Mach 1 by definition. 20 °C = 293.15 kelvins.
Muzzle velocity of a 7.62×39mm cartridge	710	2330	2600	1600	The 7.62×39mm round is a rifle cartridge of Soviet origin
Official flight airspeed record for jet engined aircraft	980	3215	3530	2194	Lockheed SR-71 Blackbird
Space shuttle on re-entry	7800	25600	28000	17,500	
Escape velocity on Earth	11200	36700	40000	25000	11.2 km·s−1
Voyager 1 relative velocity to the Sun in 2013	17000	55800	61200	38000	Fastest heliocentric recession speed of any humanmade object. (11 mi/s)
Average orbital speed of planet Earth around the Sun	29783	97713	107218	66623	
The fastest recorded speed of the Helios probes.	70,220	230,381	252,792	157,078	Recognized as the fastest speed achieved by a man-made spacecraft, achieved in solar orbit.
Speed of light in vacuum (symbol c)	299792458	983571056	1079252848	670616629	Exactly 299792458 m/s, by definition of the metre

Velocity

The velocity of an object is the rate of change of its position with respect to a frame of reference, and is a function of time. Velocity is equivalent to a specification of an object's speed and direction of motion (e.g. 60 km/h to the north). Velocity is a fundamental concept in kinematics, the branch of classical mechanics that describes the motion of bodies.

Velocity is a physical vector quantity; both magnitude and direction are needed to define it. The scalar absolute value (magnitude) of velocity is called *speed*, being a coherent derived unit whose quantity is measured in the SI (metric system) as metres per second (m/s) or as the SI base unit of ($m{\cdot}s^{-1}$). For example, "5 metres per second" is a scalar, whereas "5 metres per second east" is a vector. If there is a change in speed, direction or both, then the object has a changing velocity and is said to be undergoing an *acceleration.*

Constant Velocity vs. Acceleration

To have a constant velocity, an object must have a constant speed in a constant direction. Constant direction constrains the object to motion in a straight path thus, a constant velocity means motion in a straight line at a constant speed.

For example, a car moving at a constant 20 kilometres per hour in a circular path has a constant speed, but does not have a constant velocity because its direction changes. Hence, the car is considered to be undergoing an acceleration.

Difference between Speed and Velocity

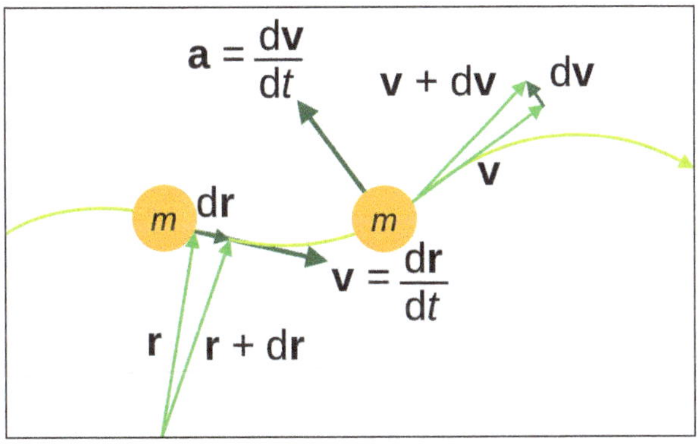

Kinematic quantities of a classical particle:
mass m, position r, velocity v, acceleration a.

Speed, the scalar magnitude of a velocity vector, denotes only how fast an object is moving.

Equation of Motion

Average Velocity

Velocity is defined as the rate of change of position with respect to time, which may also be referred to as the *instantaneous velocity* to emphasize the distinction from the average velocity. In some applications the "average velocity" of an object might be needed, that is to say, the constant velocity that would provide the same resultant displacement as a variable velocity in the same time interval, v(t), over some time period Δt. Average velocity can be calculated as:

$$\bar{v} = \frac{\Delta x}{\Delta t}$$

The average velocity is always less than or equal to the average speed of an object. This can be seen by realizing that while distance is always strictly increasing, displacement can increase or decrease in magnitude as well as change direction.

In terms of a displacement-time (x vs. t) graph, the instantaneous velocity (or, simply, velocity) can be thought of as the slope of the tangent line to the curve at any point, and the average velocity as the slope of the secant line between two points with t coordinates equal to the boundaries of the time period for the average velocity.

The average velocity is the same as the velocity averaged over time – that is to say, its time-weighted average, which may be calculated as the time integral of the velocity:

$$\bar{v} = \frac{1}{t_1 - t_0} \int_{t_0}^{t_1} v(t)\, dt,$$

where we may identify,

$$\Delta x = \int_{t_0}^{t_1} v(t)\, dt$$

and

$$\Delta t = t_1 - t_0.$$

Instantaneous Velocity

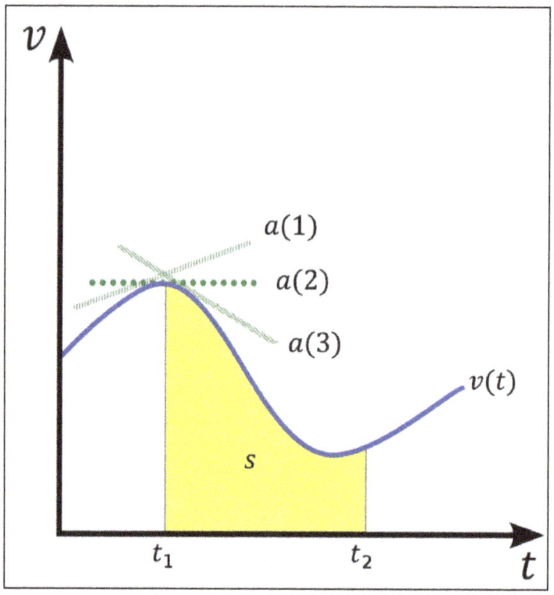

Example of a velocity vs. time graph, and the relationship between velocity v on the y-axis, acceleration a (the three green tangent lines represent the values for acceleration at different points along the curve) and displacement s (the yellow area under the curve.)

If we consider v as velocity and x as the displacement (change in position) vector, then we can express the (instantaneous) velocity of a particle or object, at any particular time t, as the derivative of the position with respect to time:

$$v = \lim_{\Delta t \to 0} \frac{\Delta x}{\Delta t} = \frac{dx}{dt}.$$

From this derivative equation, in the one-dimensional case it can be seen that the area under a velocity vs. time (v vs. t graph) is the displacement, x. In calculus terms, the integral of the velocity function $\boldsymbol{v}(t)$ is the displacement function $\boldsymbol{x}(t)$. In the figure, this corresponds to the yellow area under the curve labeled **s** (**s** being an alternative notation for displacement).

$$x = \int v \, dt.$$

Since the derivative of the position with respect to time gives the change in position (in metres) divided by the change in time (in seconds), velocity is measured in metres per second (m/s). Although the concept of an instantaneous velocity might at first seem counter-intuitive, it may be thought of as the velocity that the object would continue to travel at if it stopped accelerating at that moment.

Relationship to Acceleration

Although velocity is defined as the rate of change of position, it is often common to start with an expression for an object's acceleration. As seen by the three green tangent lines in the figure, an object's instantaneous acceleration at a point in time is the slope of the line tangent to the curve of a $\boldsymbol{v}(t)$ graph at that point. In other words, acceleration is defined as the derivative of velocity with respect to time:

$$a = \frac{dv}{dt}.$$

From there, we can obtain an expression for velocity as the area under an $\boldsymbol{a}(t)$ acceleration vs. time graph. As above, this is done using the concept of the integral:

$$v = \int a \, dt.$$

Constant Acceleration

In the special case of constant acceleration, velocity can be studied using the suvat equations. By considering a as being equal to some arbitrary constant vector, it is trivial to show that,

$$v = u + at$$

with v as the velocity at time t and u as the velocity at time $t = 0$. By combining this equation with the suvat equation $x = ut + at^2/2$, it is possible to relate the displacement and the average velocity by,

$$x = \frac{(u+v)}{2} t = \bar{v} t.$$

It is also possible to derive an expression for the velocity independent of time, known as the Torricelli equation, as follows:

$$v^2 = v \cdot v = (u + at) \cdot (u + at) = u^2 + 2t(a \cdot u) + a^2 t^2$$

$$(2a) \cdot x = (2a) \cdot (ut + \frac{1}{2} at^2) = 2t(a \cdot u) + a^2 t^2 = v^2 - u^2$$

$$\therefore v^2 = u^2 + 2(a \cdot x)$$

where $v = |v|$ etc.

The above equations are valid for both Newtonian mechanics and special relativity. Where Newtonian mechanics and special relativity differ is in how different observers would describe the same situation. In particular, in Newtonian mechanics, all observers agree on the value of t and the transformation rules for position create a situation in which all non-accelerating observers would describe the acceleration of an object with the same values. Neither is true for special relativity. In other words, only relative velocity can be calculated.

Quantities that are Dependent on Velocity

The kinetic energy of a moving object is dependent on its velocity and is given by the equation,

$$E_k = \tfrac{1}{2}mv^2$$

ignoring special relativity, where E_k is the kinetic energy and m is the mass. Kinetic energy is a scalar quantity as it depends on the square of the velocity, however a related quantity, momentum, is a vector and defined by,

$$p = mv$$

In special relativity, the dimensionless Lorentz factor appears frequently, and is given by,

$$\gamma = \frac{1}{\sqrt{1-\frac{v^2}{c^2}}}$$

where γ is the Lorentz factor and c is the speed of light.

Escape velocity is the minimum speed a ballistic object needs to escape from a massive body such as Earth. It represents the kinetic energy that, when added to the object's gravitational potential energy, (which is always negative) is equal to zero. The general formula for the escape velocity of an object at a distance r from the center of a planet with mass M is,

$$v_e = \sqrt{\frac{2GM}{r}} = \sqrt{2gr},$$

where G is the Gravitational constant and g is the Gravitational acceleration. The escape velocity from Earth's surface is about 11 200 m/s, and is irrespective of the direction of the object. This makes "escape velocity" somewhat of a misnomer, as the more correct term would be "escape speed": any object attaining a velocity of that magnitude, irrespective of atmosphere, will leave the vicinity of the base body as long as it doesn't intersect with something in its path.

Relative Velocity

Relative velocity is a measurement of velocity between two objects as determined in a single coordinate system. Relative velocity is fundamental in both classical and modern physics, since many systems in physics deal with the relative motion of two or more particles. In Newtonian mechanics, the relative velocity is independent of the chosen inertial reference frame. This is not the case anymore with special relativity in which velocities depend on the choice of reference frame.

If an object A is moving with velocity vector $\boldsymbol{v}$ and an object B with velocity vector $\boldsymbol{w}$, then the velocity of object A *relative to* object B is defined as the difference of the two velocity vectors:

$$v_{A \text{ relative to } B} = v - w$$

Similarly, the relative velocity of object B moving with velocity $\boldsymbol{w}$, relative to object A moving with velocity $\boldsymbol{v}$ is:

$$v_{B \text{ relative to } A} = w - v$$

Usually, the inertial frame chosen is that in which the latter of the two mentioned objects is in rest.

Scalar Velocities

In the one-dimensional case, the velocities are scalars and the equation is either:

$v_{rel} = v - (-w)$, if the two objects are moving in opposite directions, or:

$v_{rel} = v - (+w)$, if the two objects are moving in the same direction.

Polar Coordinates

In polar coordinates, a two-dimensional velocity is described by a radial velocity, defined as the component of velocity away from or toward the origin (also known as velocity made good), and an angular velocity, which is the rate of rotation about the origin (with positive quantities representing counter-clockwise rotation and negative quantities representing clockwise rotation, in a right-handed coordinate system).

The radial and angular velocities can be derived from the Cartesian velocity and displacement vectors by decomposing the velocity vector into radial and transverse components. The transverse velocity is the component of velocity along a circle centered at the origin.

$$v = v_T + v_R$$

where

v_T is the transverse velocity

v_R is the radial velocity.

The magnitude of the radial velocity is the dot product of the velocity vector and the unit vector in the direction of the displacement.

$$v_R = \frac{v \cdot r}{|r|}$$

where

is displacement.

The magnitude of the transverse velocity is that of the cross product of the unit vector in the direction of the displacement and the velocity vector. It is also the product of the angular speed ω and the magnitude of the displacement.

$$v_T = \frac{|r \times v|}{|r|} = \omega |r|$$

such that,

$$\omega = \frac{|r \times v|}{|r|^2}.$$

Angular momentum in scalar form is the mass times the distance to the origin times the transverse velocity, or equivalently, the mass times the distance squared times the angular speed. The sign convention for angular momentum is the same as that for angular velocity.

$$L = mrv_T = mr^2\omega$$

where

m is mass

$r = \| r \|$.

The expression mr^2 is known as moment of inertia. If forces are in the radial direction only with an inverse square dependence, as in the case of a gravitational orbit, angular momentum is constant, and transverse speed is inversely proportional to the distance, angular speed is inversely proportional to the distance squared, and the rate at which area is swept out is constant. These relations are known as Kepler's laws of planetary motion.

Acceleration

In physics, acceleration is the rate of change of velocity of an object with respect to time. An object's acceleration is the net result of all forces acting on the object, as described by Newton's Second Law. The SI unit for acceleration is metre per second squared ($m \cdot s^{-2}$). Accelerations are vector quantities (they have magnitude and direction) and add according to the parallelogram law. The vector of the net force acting on a body has the same direction as the vector of the body's acceleration, and its magnitude is proportional to the magnitude of the acceleration, with the object's mass (a scalar quantity) as proportionality constant.

For example, when a car starts from a standstill (zero velocity, in an inertial frame of reference) and travels in a straight line at increasing speeds, it is accelerating in the direction of travel. If the car turns, an acceleration occurs toward the new direction. The forward acceleration of the car is called a linear (or tangential) acceleration, the reaction to which passengers in the car experience as a force pushing them back into their seats. When changing direction, this is called radial (as orthogonal to tangential) acceleration, the reaction to which passengers experience as a sideways force. If the speed of the car decreases, this is an acceleration in the opposite direction of the velocity of the vehicle, sometimes called deceleration or Retrograde burning in spacecraft. Passengers experience the reaction to deceleration as a force pushing them forwards. Both acceleration and deceleration are treated the same, they are both changes in velocity. Each of these accelerations (tangential, radial, deceleration) is felt by passengers until their velocity (speed and direction) matches that of the uniformly moving car.

Average Acceleration

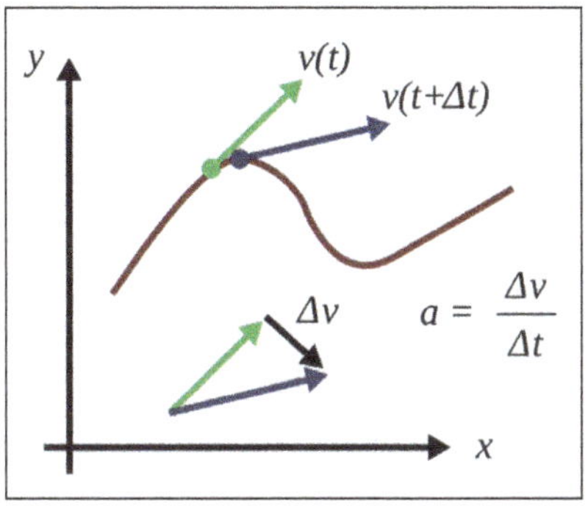

Acceleration is the rate of change of velocity. At any point on a trajectory, the magnitude of the acceleration is given by the rate of change of velocity in both magnitude and direction at that point. The true acceleration at time t is found in the limit as time interval $\Delta t \to 0$ of $\Delta v/\Delta t$.

An object's average acceleration over a period of time is its change in velocity $(\Delta \mathbf{v})$ divided by the duration of the period (Δt). Mathematically,

$$\bar{\mathbf{a}} = \frac{\Delta \mathbf{v}}{\Delta t}.$$

Instantaneous Acceleration

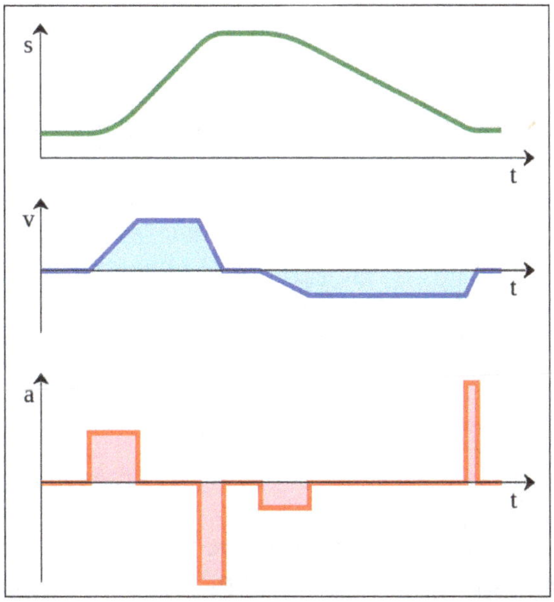

Figure: From bottom to top- (a) an acceleration function $a(t)$; (b) the integral of the acceleration is the velocity function $v(t)$; (c) and the integral of the velocity is the distance function $s(t)$.

Instantaneous acceleration, meanwhile, is the limit of the average acceleration over an infinitesimal interval of time. In the terms of calculus, instantaneous acceleration is the derivative of the velocity vector with respect to time:

$$\mathbf{a} = \lim_{\Delta t \to 0} \frac{\Delta \mathbf{v}}{\Delta t} = \frac{d\mathbf{v}}{dt}$$

(Here and elsewhere, if motion is in a straight line, vector quantities can be substituted by scalars in the equations.)

It can be seen that the integral of the acceleration function $a(t)$ is the velocity function $v(t)$; that is, the area under the curve of an acceleration vs. time (a vs. t) graph corresponds to velocity.

$$\mathbf{v} = \int \mathbf{a}\, dt$$

As acceleration is defined as the derivative of velocity, v, with respect to time t and velocity is defined as the derivative of position, x, with respect to time, acceleration can be thought of as the second derivative of x with respect to t:

$$\mathbf{a} = \frac{d\mathbf{v}}{dt} = \frac{d^2\mathbf{x}}{dt^2}$$

Units

Acceleration has the dimensions of velocity (L/T) divided by time, i.e. $L\,T^{-2}$. The SI unit of acceleration is the metre per second squared ($m\,s^{-2}$); or "metre per second per second", as the velocity in metres per second changes by the acceleration value, every second.

Other Forms

An object moving in a circular motion—such as a satellite orbiting the Earth—is accelerating due to the change of direction of motion, although its speed may be constant. In this case it is said to be undergoing *centripetal* (directed towards the center) acceleration.

Proper acceleration, the acceleration of a body relative to a free-fall condition, is measured by an instrument called an accelerometer.

In classical mechanics, for a body with constant mass, the (vector) acceleration of the body's center of mass is proportional to the net force vector (i.e. sum of all forces) acting on it (Newton's second law):

$$\mathbf{F} = m\mathbf{a} \quad \rightarrow \quad \mathbf{a} = \frac{\mathbf{F}}{m}$$

where F is the net force acting on the body, m is the mass of the body, and a is the center-of-mass acceleration. As speeds approach the speed of light, relativistic effects become increasingly large.

Tangential and Centripetal Acceleration

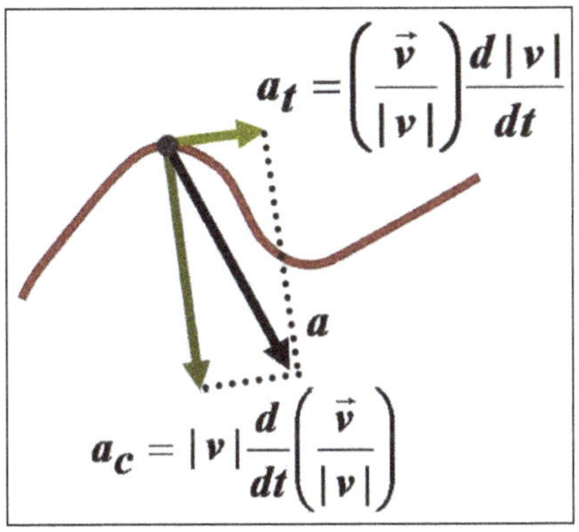

Components of acceleration for a curved motion.

The tangential component at is due to the change in speed of traversal, and points along the curve in the direction of the velocity vector (or in the opposite direction). The normal component (also called centripetal component for circular motion) ac is due to the change in direction of the velocity vector and is normal to the trajectory, pointing toward the center of curvature of the path.

The velocity of a particle moving on a curved path as a function of time can be written as:

$$\mathbf{v}(t) = v(t)\frac{\mathbf{v}(t)}{v(t)} = v(t)\mathbf{u}_t(t),$$

with $v(t)$ equal to the speed of travel along the path, and

$$\mathbf{u}_t = \frac{\mathbf{v}(t)}{v(t)},$$

a unit vector tangent to the path pointing in the direction of motion at the chosen moment in time. Taking into account both the changing speed $v(t)$ and the changing direction of u_t, the acceleration of a particle moving on a curved path can be written using the chain rule of differentiation for the product of two functions of time as:

$$\begin{aligned}\mathbf{a} &= \frac{\mathrm{d}\mathbf{v}}{\mathrm{d}t} \\ &= \frac{\mathrm{d}v}{\mathrm{d}t}\mathbf{u}_t + v(t)\frac{d\mathbf{u}_t}{dt} \\ &= \frac{\mathrm{d}v}{\mathrm{d}t}\mathbf{u}_t + \frac{v^2}{r}\mathbf{u}_n,\end{aligned}$$

where u_n is the unit (inward) normal vector to the particle's trajectory (also called the principal normal), and r is its instantaneous radius of curvature based upon the osculating circle at time t. These components are called the tangential acceleration and the normal or radial acceleration (or centripetal acceleration in circular motion).

Geometrical analysis of three-dimensional space curves, which explains tangent, (principal) normal and binormal, is described by the Frenet–Serret formulas.

Special Cases

Uniform Acceleration

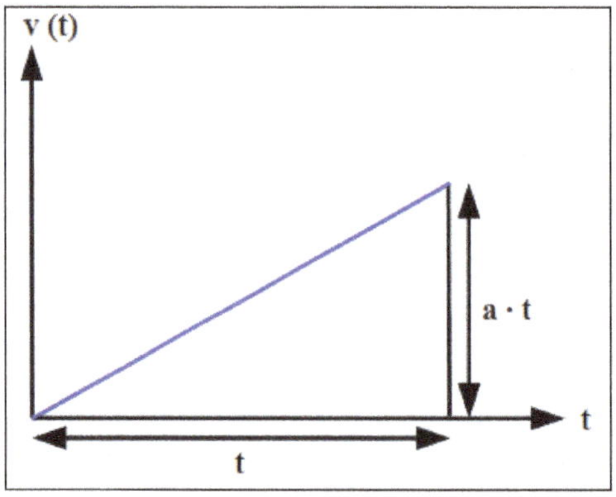

Calculation of the speed difference for a uniform acceleration.

Uniform or constant acceleration is a type of motion in which the velocity of an object changes by an equal amount in every equal time period.

A frequently cited example of uniform acceleration is that of an object in free fall in a uniform gravitational field. The acceleration of a falling body in the absence of resistances to motion is dependent only on the gravitational field strength g (also called *acceleration due to gravity*). By Newton's Second Law the force $\mathbf{F_g}$ acting on a body is given by:

$$\mathbf{F_g} = m\mathbf{g}$$

Because of the simple analytic properties of the case of constant acceleration, there are simple formulas relating the displacement, initial and time-dependent velocities, and acceleration to the time elapsed:

$$\mathbf{s}(t) = \mathbf{s}_0 + \mathbf{v}_0 t + \tfrac{1}{2}\mathbf{a}t^2 = \mathbf{s}_0 + \frac{\mathbf{v}_0 + \mathbf{v}(t)}{2}t$$

$$\mathbf{v}(t) = \mathbf{v}_0 + \mathbf{a}t$$

$$v^2(t) = v_0{}^2 + 2\mathbf{a}\cdot[\mathbf{s}(t) - \mathbf{s}_0]$$

where

t is the elapsed time,

$\mathbf{s}_0$ is the initial displacement from the origin,

$\mathbf{s}(t)$ is the displacement from the origin at time t,

$\mathbf{v}_0$ is the initial velocity,

$\mathbf{v}(t)$ is the velocity at time t, and

$\mathbf{a}$ is the uniform rate of acceleration.

In particular, the motion can be resolved into two orthogonal parts, one of constant velocity and the other according to the above equations. As Galileo showed, the net result is parabolic motion, which describes, e. g., the trajectory of a projectile in a vacuum near the surface of Earth.

Circular Motion

In uniform circular motion, that is moving with constant *speed* along a circular path, a particle experiences an acceleration resulting from the change of the direction of the velocity vector, while its magnitude remains constant. The derivative of the location of a point on a curve with respect to time, i.e. its velocity, turns out to be always exactly tangential to the curve, respectively orthogonal to the radius in this point. Since in uniform motion the velocity in the tangential direction does not change, the acceleration must be in radial direction, pointing to the center of the circle. This acceleration constantly changes the direction of the velocity to be tangent in the neighboring point, thereby rotating the velocity vector along the circle.

- For a given speed v, the magnitude of this geometrically caused acceleration (centripetal acceleration) is inversely proportional to the radius r of the circle, and increases as the square of this speed:

  $$a_c = \frac{v^2}{r}.$$

- Note that, for a given angular velocity ω, the centripetal acceleration is directly proportional to radius r. This is due to the dependence of velocity v on the radius r.

  $$v = \omega r$$

Expressing centripetal acceleration vector in polar components, where $\mathbf{r}$ is a vector from the centre of the circle to the particle with magnitude equal to this distance, and considering the orientation of the acceleration towards the center, yields

$$\mathbf{a}_c = -\frac{v^2}{|\mathbf{r}|}\cdot\frac{\mathbf{r}}{|\mathbf{r}|}.$$

As usual in rotations, the speed v of a particle may be expressed as an angular speed with respect to a point at the distance r as,

$$\omega = \frac{v}{r}.$$

Thus,

$$\mathbf{a}_c = -\omega^2\mathbf{r}.$$

This acceleration and the mass of the particle determine the necessary centripetal force, directed *toward* the centre of the circle, as the net force acting on this particle to keep it in this uniform circular motion. The so-called 'centrifugal force', appearing to act outward on the body, is a so-called pseudo force experienced in the frame of reference of the body in circular motion, due to the body's linear momentum, a vector tangent to the circle of motion.

In a nonuniform circular motion, i.e., the speed along the curved path is changing, the acceleration has a non-zero component tangential to the curve, and is not confined to the principal normal, which directs to the center of the osculating circle, that determines the radius r for the centripetal acceleration. The tangential component is given by the angular acceleration α, i.e., the rate of change $\alpha = \dot{\omega}$ of the angular speed ω times the radius r. That is,

$$a_c = r\alpha.$$

The sign of the tangential component of the acceleration is determined by the sign of the angular acceleration (α), and the tangent is of course always directed at right angles to the radius vector.

Relation to Relativity

Special Relativity

The special theory of relativity describes the behavior of objects traveling relative to other objects at speeds approaching that of light in a vacuum. Newtonian mechanics is exactly revealed to be an approximation to reality, valid to great accuracy at lower speeds. As the relevant speeds increase toward the speed of light, acceleration no longer follows classical equations.

As speeds approach that of light, the acceleration produced by a given force decreases, becoming infinitesimally small as light speed is approached; an object with mass can approach this speed asymptotically, but never reach it.

General Relativity

Unless the state of motion of an object is known, it is impossible to distinguish whether an observed force is due to gravity or to acceleration—gravity and inertial acceleration have identical effects. Albert Einstein called this the equivalence principle, and said that only observers who feel no force at all—including the force of gravity—are justified in concluding that they are not accelerating.

Conversions

Conversions between common units of acceleration				
Base value	(Gal, or cm/s²)	(ft/s²)	(m/s²)	(Standard gravity, g_0)
1 Gal, or cm/s²	1	0.0328084	0.01	0.00101972
1 ft/s²	30.4800	1	0.304800	0.0310810
1 m/s²	100	3.28084	1	0.101972
1 g_0	980.665	32.1740	9.80665	1

Motion

In physics, motion is the change with time of the position or orientation of a body. Motion along a line or a curve is called translation. Motion that changes the orientation of a body is called rotation.

In both cases all points in the body have the same velocity (directed speed) and the same acceleration (time rate of change of velocity). The most general kind of motion combines both translation and rotation.

All motions are relative to some frame of reference. Saying that a body is at rest, which means that it is not in motion, merely means that it is being described with respect to a frame of reference that is moving together with the body. For example, a body on the surface of the Earth may appear to be at rest, but that is only because the observer is also on the surface of the Earth. The Earth itself, together with both the body and the observer, is moving in its orbit around the Sun and rotating on its own axis at all times. As a rule, the motions of bodies obey Newton's laws of motion. However, motion at speeds close to the speed of light must be treated by using the theory of relativity, and the motion of very small bodies (such as electrons) must be treated by using quantum mechanics.

Motion of a Particle in one Dimension

Uniform Motion

According to Newton's first law (also known as the principle of inertia), a body with no net force acting on it will either remain at rest or continue to move with uniform speed in a straight line, according to its initial condition of motion. In fact, in classical Newtonian mechanics, there is no important distinction between rest and uniform motion in a straight line; they may be regarded as the same state of motion seen by different observers, one moving at the same velocity as the particle, the other moving at constant velocity with respect to the particle.

Although the principle of inertia is the starting point and the fundamental assumption of classical mechanics, it is less than intuitively obvious to the untrained eye. In Aristotelian mechanics, and in ordinary experience, objects that are not being pushed tend to come to rest. The law of inertia was deduced by Galileo from his experiments with balls rolling down inclined planes.

For Galileo, the principle of inertia was fundamental to his central scientific task: he had to explain how it is possible that if Earth is really spinning on its axis and orbiting the Sun we do not sense that motion. The principle of inertia helps to provide the answer: Since we are in motion together with Earth, and our natural tendency is to retain that motion, Earth appears to us to be at rest. Thus, the principle of inertia, far from being a statement of the obvious, was once a central issue of scientific contention. By the time Newton had sorted out all the details, it was possible to account accurately for the small deviations from this picture caused by the fact that the motion of Earth's surface is not uniform motion in a straight line. In the Newtonian formulation, the common observation that bodies that are not pushed tend to come to rest is attributed to the fact that they have unbalanced forces acting on them, such as friction and air resistance.

As has already been stated, a body in motion may be said to have momentum equal to the product of its mass and its velocity. It also has a kind of energy that is due entirely to its motion, called kinetic energy. The kinetic energy of a body of mass m in motion with velocity $\boldsymbol{v}$ is given by,

$$K = \frac{1}{2}mv^2$$

Falling Bodies and Uniformly Accelerated Motion

During the 14th century, the French scholar Nicole Oresme studied the mathematical properties of uniformly accelerated motion. He had little interest in whether that kind of motion could be observed in the realm of actual human existence, but he did discover that, if a particle is uniformly accelerated, its speed increases in direct proportion to time, and the distance it traverses is proportional to the square of the time spent accelerating. Two centuries later, Galileo repeated these same mathematical discoveries (perhaps independently) and, just as important, determined that this kind of motion is actually executed by balls rolling down inclined planes. As the incline of the plane increases, the acceleration increases, but the motion continues to be uniformly accelerated. From this observation, Galileo deduced that a body falling freely in the vertical direction would also have uniform acceleration. Even more remarkably, he demonstrated that, in the absence of air resistance, all bodies would fall with the same constant acceleration regardless of their mass. If the constant acceleration of any body dropped near the surface of Earth is expressed as g, the behaviour of a body dropped from rest at height z_0 and time $t = 0$ may be summarized by the following equations:

$$z = z_0 - \frac{1}{2}gt^2,$$

$$v = gt,$$

$$a = g$$

where z is the height of the body above the surface, v is its speed, and a is its acceleration. These equations of motion hold true until the body actually strikes the surface. The value of g is approximately 9.8 metres per second squared (m/s^2).

A body of mass m at a height z_0 above the surface may be said to possess a kind of energy purely by virtue of its position. This kind of energy (energy of position) is called potential energy. The gravitational potential energy is given by,

$$U = mgz_0.$$

Technically, it is more correct to say that this potential energy is a property of the Earth-body system rather than a property of the body itself, but this pedantic distinction can be ignored.

As the body falls to height z less than z_0, its potential energy U converts to kinetic energy $K = {}^1/_2mv^2$. Thus, the speed v of the body at any height z is given by solving the equation,

$$\frac{1}{2}mv^2 + mgz = mgz_0.$$

Equation above is an expression of the law of conservation of energy. It says that the sum of kinetic energy, ${}^1/_2mv^2$, and potential energy, mgz, at any point during the fall, is equal to the total initial energy, mgz_0, before the fall began. Exactly the same dependence of speed on height could be deduced from the kinematic equations $z = z_0 - \frac{1}{2}gt^2$, $v = gt$, and $a = g$ above.

In order to reach the initial height z_0, the body had to be given its initial potential energy by some external agency, such as a person lifting it. The process by which a body or a system obtains mechanical energy from outside of itself is called work. The increase of the energy of the body is equal to the work done on it. Work is equal to force times distance.

The force exerted by Earth's gravity on a body of mass m may be deduced from the observation that the body, if released, will fall with acceleration g. Since force is equal to mass times acceleration, the force of gravity is given by $F = mg$. To lift the body to height z_0, an equal and opposite (i.e., upward) force must be exerted through a distance z_0. Thus, the work done is,

$$W = Fz_0 = mgz_0,$$

which is equal to the potential energy that results.

If work is done by applying a force to a body that is not being acted upon by an opposing force, the body is accelerated. In this case, the work endows the body with kinetic energy rather than potential energy. The energy that the body gains is equal to the work done on it in either case. It should be noted that work, potential energy, and kinetic energy, all being aspects of the same quantity, must all have the dimensions ml^2/t^2.

Simple Harmonic Oscillations

Consider a mass m held in an equilibrium position by springs, as shown in Figure. The mass may be perturbed by displacing it to the right or left. If x is the displacement of the mass from equilibrium, the springs exert a force F proportional to x, such that

$$F = -kx,$$

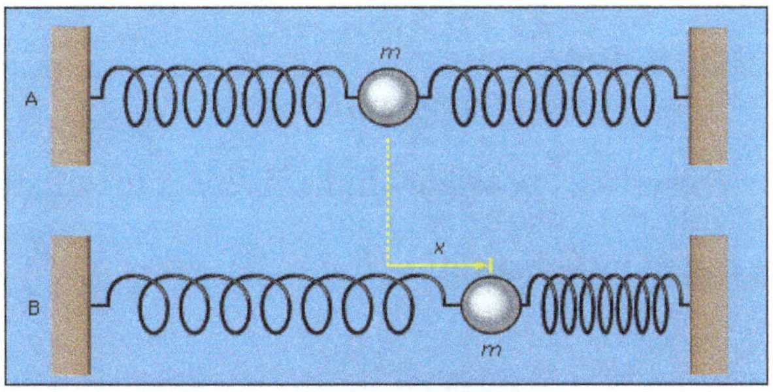

Figure: (A) A mass m held in equilibrium by springs.
(B) A mass m displaced a distance x.

where k is a constant that depends on the stiffness of the springs. Equation $F = -kx$, is called Hooke's law, and the force is called the spring force. If x is positive (displacement to the right), the resulting force is negative (to the left), and vice versa. In other words, the spring force always acts so as to restore mass back toward its equilibrium position. Moreover, the force will produce an acceleration along the x direction given by $a = d^2x/dt^2$. Thus, Newton's second law, $F = ma$, is applied to this case by substituting $-kx$ for F and d^2x/dt^2 for a, giving $-kx = m(d^2x/dt^2)$. Transposing and dividing by m yields the equation,

$$a = \frac{d^2x}{dt^2} = -\frac{k}{m}x.$$

Equation $a = \frac{d^2x}{dt^2} = -\frac{k}{m}x$ gives the derivative—in this case the second derivative—of a quantity x in terms of the quantity itself. Such an equation is called a differential equation, meaning an equation containing derivatives. Much of the ordinary, day-to-day work of theoretical physics consists of solving differential equations. The question is, given equation $a = \frac{d^2x}{dt^2} = -\frac{k}{m}x.$, how does x depend on time?

The answer is suggested by experience. If the mass is displaced and released, it will oscillate back and forth about its equilibrium position. That is, x should be an oscillating function of t, such as a sine wave or a cosine wave. For example, x might obey a behaviour such as,

$$x = A\cos\omega t.$$

Equation $x = A\cos\omega t.$ describes the behaviour sketched graphically in Figure . The mass is initially displaced a distance $x = A$ and released at time $t = 0$. As time goes on, the mass oscillates from A to $-A$ and back to A again in the time it takes ωt to advance by 2π. This time is called T, the period of oscillation, so that $\omega T = 2\pi$, or $T = 2\pi/\omega$. The reciprocal of the period, or the frequency f, in oscillations per second, is given by $f = 1/T = \omega/2\pi$. The quantity ω is called the angular frequency and is expressed in radians per second.

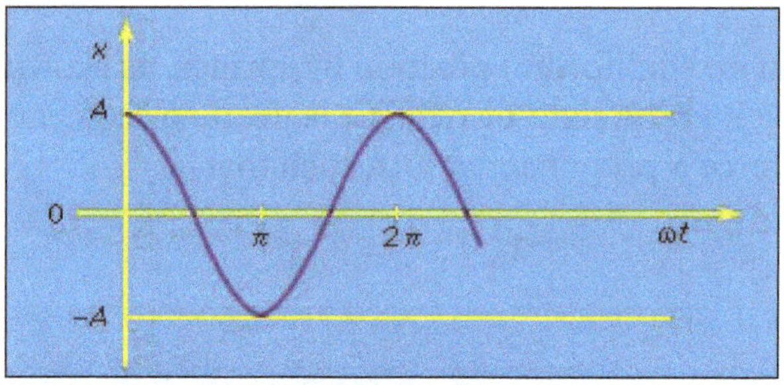

Figure: The function $x = A\cos\omega t$.

The choice of equation $x = A\cos\omega t$ as a possible kind of behaviour satisfying the differential equation $a = \frac{d^2x}{dt^2} = -\frac{k}{m}x.$ can be tested by substituting it into equation $a = \frac{d^2x}{dt^2} = -\frac{k}{m}x.$ The first derivative of x with respect to t is,

$$\begin{aligned}\frac{dx}{dt} &= \frac{d}{dt}\left(A\cos\omega t\right)\\ &= -\omega A\sin\omega t.\end{aligned}$$

Differentiating a second time gives,

$$\begin{aligned}\frac{d^2x}{dt^2} &= \frac{d}{dt}\left(\frac{dx}{dt}\right)\\ &= \frac{d}{dt}(-\omega A\sin\omega t)\\ &= -\omega^2 A\cos\omega t\\ &= -\omega^2 x.\end{aligned}$$

Equation above is the same as equation $a = \frac{d^2x}{dt^2} = -\frac{k}{m}x$ if,

$$\omega^2 = \frac{k}{m}.$$

Thus, subject to this condition, equation $x = A\cos\omega t$ is a correct solution to the differential equation. There are other possible correct guesses (e.g., $x = A \sin \omega t$) that differ from this one only in whether the mass is at rest or in motion at the instant $t = 0$.

The mass, as has been shown, oscillates from A to $-A$ and back again. The speed, given by dx/dt, equation

$$\frac{dx}{dt} = \frac{d}{dt}(A\cos\omega t)$$
$$= -\omega A \sin\omega t.$$

is zero at A and $-A$, but has its maximum magnitude, equal to ωA, when x is equal to zero. Physically, after the mass is displaced from equilibrium a distance A to the right, the restoring force F pushes the mass back toward its equilibrium position, causing it to accelerate to the left. When it reaches equilibrium, there is no force acting on it at that instant, but it is moving at speed ωA, and its inertia takes it past the equilibrium position. Before it is stopped it reaches position $-A$, and by this time there is a force acting on it again, pushing it back toward equilibrium.

The whole process, known as simple harmonic motion, repeats itself endlessly with a frequency given by equation $\omega^2 = \frac{k}{m}$. Equation $\omega^2 = \frac{k}{m}$ means that the stiffer the springs (i.e., the larger k), the higher the frequency (the faster the oscillations). Making the mass greater has exactly the opposite effect, slowing things down.

One of the most important features of harmonic motion is the fact that the frequency of the motion, ω (or f), depends only on the mass and the stiffness of the spring. It does not depend on the amplitude A of the motion. If the amplitude is increased, the mass moves faster, but the time required for a complete round trip remains the same. This fact has profound consequences, governing the nature of music and the principle of accurate timekeeping.

The potential energy of a harmonic oscillator, equal to the work an outside agent must do to push the mass from zero to x, is $U = {}^1/_2kx^2$. Thus, the total initial energy in the situation described above is ${}^1/_2kA^2$; and since the kinetic energy is always ${}^1/_2mv^2$, when the mass is at any point x in the oscillation,

$$\frac{1}{2}mv^2 + \frac{1}{2}kx^2 = \frac{1}{2}kA^2.$$

Equation above plays exactly the role for harmonic oscillators that equation,

$\frac{1}{2}mv^2 + mgz = mgz_0.$ does for falling bodies.

It is quite generally true that harmonic oscillations result from disturbing any body or structure from a state of stable mechanical equilibrium. To understand this point.

Consider a bowl with a marble resting inside, then consider a second, inverted bowl with a marble balanced on top. In both cases, the net force on the marble is zero. The marbles are thus in mechanical equilibrium. However, a small disturbance in the position of the marble balanced on top of the inverted bowl will cause it to roll away and not return. In such a case, the equilibrium is said to be unstable. Conversely, if the marble inside the first bowl is disturbed, gravity acts to push it back toward the bottom of the bowl. The marble inside the bowl (like the mass held by springs in figure) is an example of a body in stable equilibrium. If it is disturbed slightly, it executes harmonic oscillations around the bottom of the bowl rather than rolling away.

This argument may be generalized by a simple mathematical argument. Consider a body or structure in mechanical equilibrium, which, when disturbed by a small amount x, finds a force acting on it that is a function of x, $F(x)$. For small x, such a function may be written generally as a power series in x; i.e.,

$$F(x) = F(0) + ax + bx^2 + ...,$$

where $F(0)$ is the value of $F(x)$ when $x = (0)$, and a and b are constants, independent of x, determined by the nature of the system. The statement that the body is in mechanical equilibrium means that $F(0) = 0$, so that no force is acting on the body when it is undisturbed. Since x is small, x^2 is much smaller; thus, the term bx^2 and all higher powers may be disregarded. This leaves $F(x) = ax$. Now, if a is positive, a disturbance produces a force in the same direction as the disturbance. This was the case when the marble was balanced on top of the inverted bowl. It describes unstable equilibrium. For the system to be stable, a must be negative. Thus, if $a = -k$, where k is some positive constant, equation $F(x) = F(0) + ax + bx^2 + ...$, becomes $F(x) = -kx$, which is simply Hooke's law, equation $F = -kx$, As has been described above, any system obeying Hooke's law is a harmonic oscillator.

The generality of this argument accounts for the fact that harmonic oscillators are abundantly observed in common experience. For example, any rigid structure will oscillate at many different harmonic frequencies corresponding to different possible distortions of its equilibrium shape. In addition, music may be produced either by disturbing the equilibrium of a stretched wire or fibre (as in the piano and violin), a stretched membrane (e.g., drums), or a rigid bar (the triangle and the xylophone) or by disturbing the density of an enclosed column of air (as in the trumpet and organ). While a fluid such as air is not rigid, its density is an example of a stable system that obeys Hooke's law and may therefore be set into harmonic oscillations.

All music would be quite different from what it is were it not for the general property of harmonic oscillators that the frequency is independent of the amplitude. Thus, instruments yield the same note (frequency) regardless of how loudly they are played (amplitude), and, equally important, the same note persists as the vibrations die away. This same property of harmonic oscillators is the underlying principle of all accurate timekeeping.

The first precise timekeeping mechanism, whose principles of motion were discovered by Galileo, was the simple pendulum. The accuracy of modern timekeeping has been improved dramatically by the introduction of tiny quartz crystals, whose harmonic oscillations generate electrical signals that may be incorporated into miniaturized circuits in clocks and wristwatches. All harmonic oscillators are natural timekeeping devices because they oscillate at intrinsic natural frequencies independent of amplitude. A given number of complete cycles always corresponds to the same elapsed time. Quartz crystal oscillators make more accurate clocks than pendulums do principally because they oscillate many more times per second.

Damped and Forced Oscillations

The simple harmonic oscillations continue forever, at constant amplitude, oscillating as shown in figure between A and $-A$. Common experience indicates that real oscillators behave somewhat differently, however. Harmonic oscillations tend to die away as time goes on. This behavior, called damping of the oscillations, is produced by forces such as friction and viscosity. These forces are known collectively as dissipative forces because they tend to dissipate the potential and kinetic energies of macroscopic bodies into the energy of the chaotic motion of atoms and molecules known as heat.

Friction and viscosity are complicated phenomena whose effects cannot be represented accurately by a general equation. However, for slowly moving bodies, the dissipative forces may be represented by,

$$F_d = -\gamma v,$$

where v is the speed of the body and γ is a constant coefficient, independent of dynamic quantities such as speed or displacement. Equation $F_d = -\gamma v$, is most easily understood by an argument analogous to that applied to equation $F(x) = F(0) + ax + bx^2 + ...$, above. F_d is written as a sum of powers of v, or $F_d(v) = F_d(0) + av + bv^2 + \cdot\cdot\cdot$. When the body is at rest ($v = 0$), no dissipative force is expected because, if there were one, it might set the body into motion. Thus, $F_d(0) = 0$. The next term must be negative since dissipative forces always resist the motion. Thus, $a = -\gamma$ where γ is positive. Since v^2 has the same sign regardless of the direction of the motion, b must equal 0 lest it sometimes contribute a dissipative force in the same direction as the motion. The next term is proportional to v^3, and it and all subsequent terms may be neglected if v is sufficiently small. So, as in equation $F(x) = F(0) + ax + bx^2 + ...$, the power series is reduced to a single term, in this case $F_d = -\gamma v$.

To find the effect of a dissipative force on a harmonic oscillator, a new differential equation must be solved. The net force, or mass times acceleration, written as md^2x/dt^2, is set equal to the sum of the Hooke's law force, $-kx$, and the dissipative force, $-\gamma v = -\gamma dx/dt$. Dividing by m yields,

$$\frac{d^2x}{dt^2} = -\frac{k}{m}x - \frac{\gamma}{m}\frac{dx}{dt}.$$

The general solution to equation $\frac{d^2x}{dt^2} = -\frac{k}{m}x - \frac{\gamma}{m}\frac{dx}{dt}$ is given in the form $x = Ce^{-\gamma t/2m}\cos(\omega t + \theta_0)$, where C and θ_0 are arbitrary constants determined by the initial conditions. This motion, for the case in which $\theta_0 = 0$, is illustrated in figure. As expected, the harmonic oscillations die out with timc. The amplitude of the oscillations is bounded by an exponentially decreasing function of time (the dashed curves). The characteristic decay time (after which the oscillations are smaller by $1/e$, where e is the base of the natural logarithms $e = 2.718\ldots$) is equal to $2m/\gamma$. The frequency of the oscillations is given by:

$$\omega^2 = \frac{k}{m} - \frac{\gamma^2}{4m^2}$$

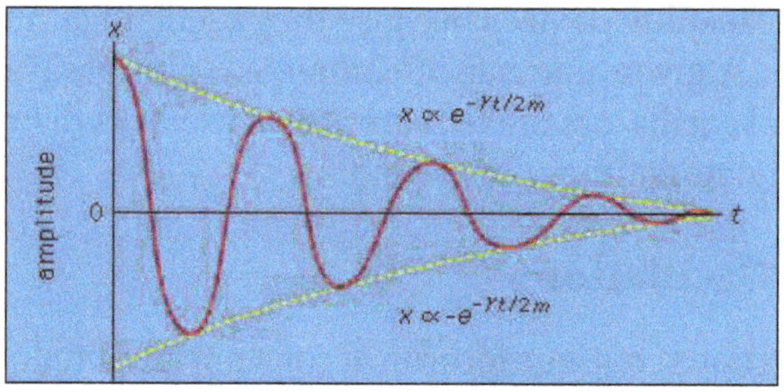

Figure: Damped oscillations.

Importantly, this frequency does not change as the oscillations decay.

Equation $\omega^2 = \frac{k}{m} - \frac{\gamma^2}{4m^2}$ shows that it is possible, by proper choice of γ, to turn a harmonic oscillator into a system that does not oscillate at all—that is, a system whose natural frequency is ω = 0. Such a system is said to be critically damped. For example, the springs that suspend the body of an automobile cause it to be a natural harmonic oscillator. The shock absorbers of the auto are devices that seek to add just enough dissipative force to make the assembly critically damped. In this way, the passengers need not go through numerous oscillations after each bump in the road.

$$\omega^2 = \frac{k}{m} - \frac{\gamma^2}{4m^2}$$

A simple disturbance can set a harmonic oscillator into motion. Repeated disturbances can increase the amplitude of the oscillations if they are applied in synchrony with the natural frequency. Even a very small disturbance, repeated periodically at just the right frequency, can cause a very large amplitude motion to build up. This phenomenon is known as resonance.

Periodically forced oscillations may be represented mathematically by adding a term of the form a_0 sin ω*t* to the right-hand side of equation $\frac{d^2x}{dt^2} = -\frac{k}{m}x - \frac{\gamma}{m}\frac{dx}{dt}$. This term describes a force applied at frequency ω, with amplitude ma_0. The result of applying such a force is to create a kind of motion that does not need to decay with time, since the energy lost to dissipative processes is replaced, over the course of each cycle, by the driving force. The amplitude of the motion depends on how close the driving frequency ω is to the natural frequency ω_0 of the oscillator. Interestingly, even though dissipation is present, ω_0 is not given by equation above but rather by equation $\omega^2 = \frac{k}{m}$: $\omega^2_0 = k/m$. In a graph of the amplitude of the steady state motion (i.e., long after the driving force has begun to be applied), the maximum amplitude occurs as expected at $\omega = \omega_0$. The height and width of the resonance curve are governed by the damping coefficient γ. If there were no damping, the maximum amplitude would be infinite. Because small disturbances at every possible frequency are always present in the natural world, every rigid structure would shake itself to pieces if not for the presence of internal damping.

Resonances are not uncommon in the world of familiar experience. For example, cars often rattle at certain engine speeds, and windows sometimes rattle when an airplane flies by. Resonance is particularly important in music. For example, the sound box of a violin does its job well if it has a natural frequency of oscillation that responds resonantly to each musical note. Very strong resonances to certain notes—called "wolf notes" by musicians—occur in cheap violins and are much to be avoided. Sometimes, a glass may be broken by a singer as a result of its resonant response to a particular musical note.

Motion of a Particle in Two or more Dimensions

Projectile Motion

Galileo was quoted above pointing out with some detectable pride that none before him had realized that the curved path followed by a missile or projectile is a parabola. He had arrived at his conclusion by realizing that a body undergoing ballistic motion executes, quite independently, the motion of a freely falling body in the vertical direction and inertial motion in the horizontal direction. These considerations, and terms such as ballistic and projectile, apply to a body that, once launched, is acted upon by no force other than Earth's gravity.

Projectile motion may be thought of as an example of motion in space—that is to say, of three-dimensional motion rather than motion along a line, or one-dimensional motion. In a suitably defined system of Cartesian coordinates, the position of the projectile at any instant may be specified by giving the values of its three coordinates, $x(t)$, $y(t)$, and $z(t)$. By generally accepted convention, $z(t)$ is used to describe the vertical direction. To a very good approximation, the motion is confined to a single vertical plane, so that for any single projectile it is possible to choose a coordinate system such that the motion is two-dimensional [say, $x(t)$ and $z(t)$] rather than three-dimensional [$x(t)$, $y(t)$, and $z(t)$]. It is assumed that the range of the motion is sufficiently limited that the curvature of Earth's surface may be ignored.

Consider a body whose vertical motion obeys equation $z = z_0 - \frac{1}{2}gt^2$, Galileo's law of falling bodies, which states $z = z_0 - {}^1/_2 gt^2$, while, at the same time, moving horizontally at a constant speed v_x in accordance with Galileo's law of inertia. The body's horizontal motion is thus described by $x(t) = v_x t$, which may be written in the form $t = x/v_x$. Using this result to eliminate t from equation $z = z_0 - \frac{1}{2}gt^2$, gives $z = z_0 - {}^1/_2 g(1/vx)^2 x^2$. This latter is the equation of the trajectory of a projectile in the z–x plane, fired horizontally from an initial height z_0. It has the general form,

$$z = a + bx^2,$$

where a and b are constants. Equation $z = a + bx^2$, may be recognized to describe a parabola just as Galileo claimed. The parabolic shape of the trajectory is preserved even if the motion has an initial component of velocity in the vertical direction.

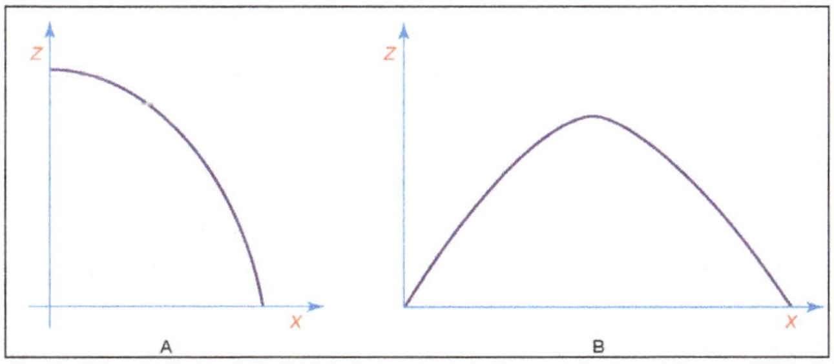

Figure: (A) The parabolic path of a projectile. (B) The parabolic path of a projectile with an initial upward component of velocity.

Energy is conserved in projectile motion. The potential energy $U(z)$ of the projectile is given by $U(z) = mgz$. The kinetic energy K is given by $K = {}^1/_2 mv^2$, where v^2 is equal to the sum of the squares of the vertical and horizontal components of velocity, or $v^2 = v^2_x + v^2_z$.

The effects of air resistance (to say nothing of wind and other more complicated phenomena) have been neglected. These effects are seldom actually negligible. They are most nearly so for bodies that are heavy and slow-moving. All of this discussion, therefore, is of great value for understanding the underlying principles of projectile motion but of little utility for predicting the actual trajectory of, say, a cannonball once fired or even a well-hit baseball.

Motion of a Pendulum

According to legend, Galileo discovered the principle of the pendulum while attending mass at the Duomo (cathedral) located in the Piazza del Duomo of Pisa, Italy. A lamp hung from the ceiling by a cable and, having just been lit, was swaying back and forth. Galileo realized that each complete cycle of the lamp took the same amount of time, compared to his own pulse, even though the amplitude of each swing was smaller than the last. As has already been shown, this property is common to all harmonic oscillators, and, indeed, Galileo's discovery led directly to the invention of the first accurate mechanical clocks. Galileo was also able to show that the period of oscillation of a simple pendulum is proportional to the square root of its length and does not depend on its mass.

A simple pendulum is sketched in Figure . A bob of mass M is suspended by a massless cable or bar of length L from a point about which it pivots freely. The angle between the cable and the vertical is called θ. The force of gravity acting on the mass M, always equal to $-Mg$ in the vertical direction, is a vector that may be resolved into two components, one that acts ineffectually along the cable and another, perpendicular to the cable, that tends to restore the bob to its equilibrium position directly below the point of suspension. This latter component is given by,

$$F = -Mg \sin\theta.$$

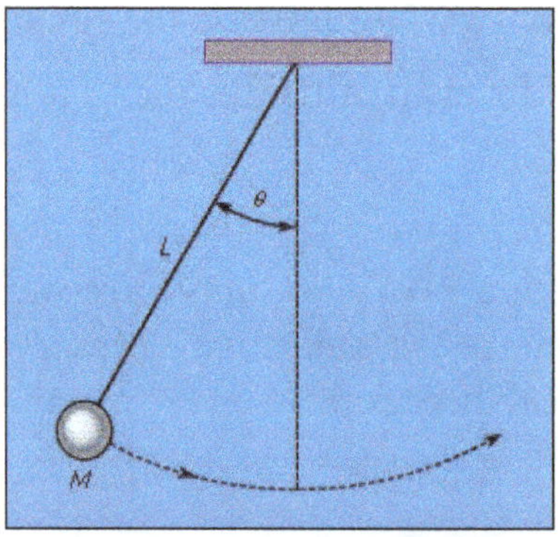

Figure: A simple pendulum.

The bob is constrained by the cable to swing through an arc that is actually a segment of a circle of radius L. If the cable is displaced through an angle θ, the bob moves a distance $L\theta$ along its arc (θ must be expressed in radians for this form to be correct). Thus, Newton's second law may be written,

$$F = Ma = M\frac{d^2(L\theta)}{dt^2}$$

Equating equation $F = -Mg\sin\theta$ to equation $F = Ma = M\frac{d^2(L\theta)}{dt^2}$, one sees immediately that the mass M will drop out of the resulting equation. The simple pendulum is an example of a falling body, and its dynamics do not depend on its mass for exactly the same reason that the acceleration of a falling body does not depend on its mass: both the force of gravity and the inertia of the body are proportional to the same mass, and the effects cancel one another. The equation that results (after extracting the constant L from the derivative and dividing both sides by L) is,

$$\frac{d^2}{dt^2} = -\frac{g}{L}\sin\theta.$$

If the angle θ is sufficiently small, equation $\frac{d^2}{dt^2} = -\frac{g}{L}\sin\theta$ may be rewritten in a form that is both more familiar and more amenable to solution. Figure shows a segment of a circle of radius L. A radius vector at angle θ, as shown, locates a point on the circle displaced a distance $L\theta$ along the arc. It is clear from the geometry that $L \sin\theta$ and $L\theta$ are very nearly equal for small θ. It follows then that sin θ and θ are also very nearly equal for small θ. Thus, if the analysis is restricted to small angles, then sin θ may be replaced by θ in equation $\frac{d^2}{dt^2} = -\frac{g}{L}\sin\theta.$ to obtain,

$$\frac{d^2\theta}{dt^2} = -\frac{g}{L}\theta.$$

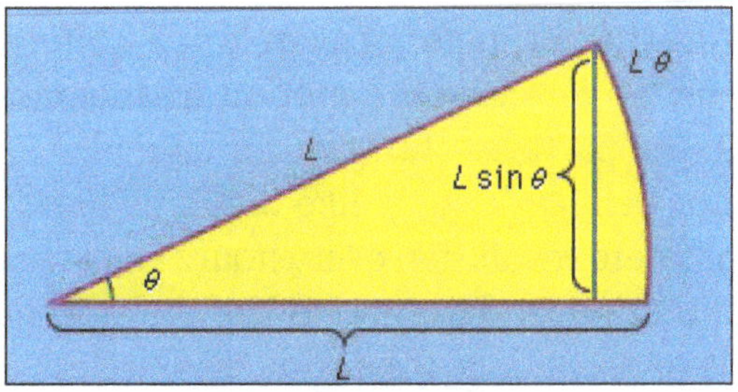

Figure: A segment of a circle of radius.

Equation $\frac{d^2\theta}{dt^2} = -\frac{g}{L}\theta$ should be compared with equation $a = \frac{d^2x}{dt^2} = -\frac{k}{m}x$: $d^2x/dt^2 = -(k/m)x$. In the first case, the dynamic variable (meaning the quantity that changes with time) is θ, in the second case it is x. In both cases, the second derivative of the dynamic variable with respect to time is equal to the variable itself multiplied by a negative constant. The equations are therefore mathematically identical and have the same solution—i.e., equation $x = A\cos\omega t$, or $\theta = A \cos \omega t$. In the case of the pendulum, the frequency of the oscillations is given by the constant in equation $\frac{d^2\theta}{dt^2} = -\frac{g}{L}\theta$, or $\omega^2 = g/L$. The period of oscillation, $T = 2\pi/\omega$, is therefore,

$$T = 2\pi\sqrt{\frac{L}{g}}.$$

Just as Galileo concluded, the period is independent of the mass and proportional to the square root of the length.

As with most problems in physics, this discussion of the pendulum has involved a number of simplifications and approximations. Most obviously, sin θ was replaced by θ to obtain equation $\frac{d^2\theta}{dt^2} = -\frac{g}{L}\theta$. This approximation is surprisingly accurate. For example, at a not-very-small angle of 17.2°, corresponding to 0.300 radian, sin θ is equal to 0.296, an error of less than 2 percent. For smaller angles, of course, the error is appreciably smaller.

The problem was also treated as if all the mass of the pendulum were concentrated at a point at the end of the cable. This approximation assumes that the mass of the bob at the end of the cable is much larger than that of the cable and that the physical size of the bob is small compared with the length of the cable. When these approximations are not sufficient, one must take into account the way in which mass is distributed in the cable and bob. This is called the physical pendulum, as opposed to the idealized model of the simple pendulum. Significantly, the period of a physical pendulum does not depend on its total mass either.

The effects of friction, air resistance, and the like have also been ignored. These dissipative forces have the same effects on the pendulum as they do on any other kind of harmonic oscillator They cause the amplitude of a freely swinging pendulum to grow smaller on successive swings. Conversely, in order to keep a pendulum clock going, a mechanism is needed to restore the energy lost to dissipative forces.

Circular Motion

Consider a particle moving along the perimeter of a circle at a uniform rate, such that it makes one complete revolution every hour. To describe the motion mathematically, a vector is constructed from the centre of the circle to the particle. The vector then makes one complete revolution every hour. In other words, the vector behaves exactly like the large hand on a wristwatch, an arrow of fixed length that makes one complete revolution every hour. The motion of the point of the vector is an example of uniform circular motion, and the period *T* of the motion is equal to one hour ($T = 1$ h). The arrow sweeps out an angle of 2π radians (one complete circle) per hour. This rate is called the angular frequency and is written $\omega = 2\pi\ h^{-1}$. Quite generally, for uniform circular motion at any rate,

$$T = \frac{2\pi}{\omega}.$$

These definitions and relations are the same as they are for harmonic motion.

Consider a coordinate system, as shown in figure below A, with the circle centred at the origin. At any instant of time, the position of the particle may be specified by giving the radius *r* of the circle and the angle θ between the position vector and the *x*-axis. Although *r* is constant, θ increases uniformly with time *t*, such that $\theta = \omega t$, or $d\theta/dt = \omega$, where ω is the angular frequency in equation $T = \frac{2\pi}{\omega}$. Contrary to the case of the wristwatch, however, ω is positive by convention when the rotation is in the counterclockwise sense. The vector *r* has *x* and *y* components given by

$$T = \frac{2\pi}{\omega}$$

$$x = r\cos\theta = r\cos\omega t,$$

$$y = r\sin\theta = r\sin\omega t.$$

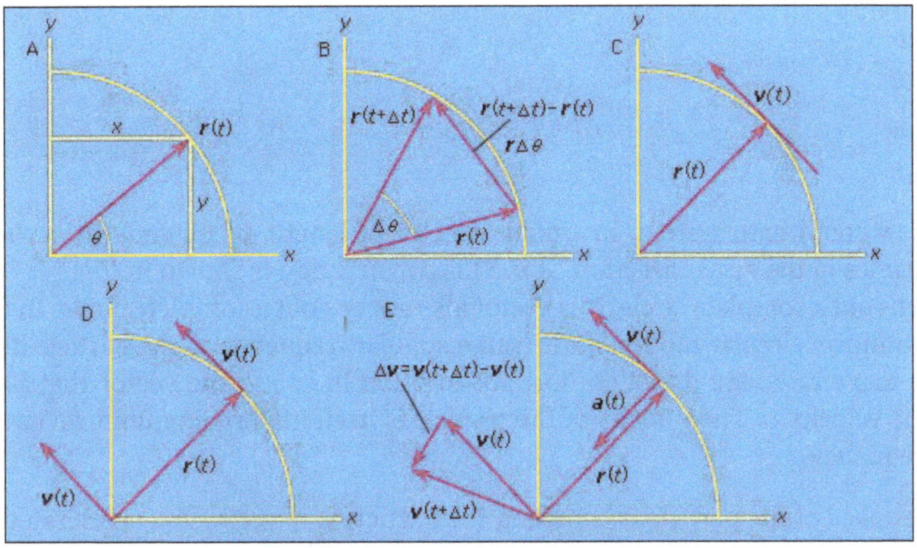

In the figure above, (A) A coordinate system to describe uniform circular motion. (B) The distance traveled in time Δt by a particle undergoing uniform circular motion. (C) The instantaneous velocity of the particle. (D) The velocity vector v undergoes uniform circular motion at the same angular frequency as the particle. (E) The acceleration vector of the particle.

One meaning of equations $x = r\cos\theta = r\cos\omega t$, and $y = r\sin\theta = r\sin\omega t$, is that, when a particle undergoes uniform circular motion, its x and y components each undergo simple harmonic motion. They are, however, not in phase with one another: at the instant when x has its maximum amplitude (say, at $\theta = 0$), y has zero amplitude, and vice versa.

In a short time, Δt, the particle moves $r\Delta\theta$ along the circumference of the circle, as shown in figure B. The average speed of the particle is thus given by,

$$\bar{v} = r\frac{\Delta\theta}{\Delta t}.$$

The average velocity of the particle is a vector given by,

$$\bar{v} = \frac{r(t+\Delta t) - r(t)}{\Delta t}.$$

This operation of vector subtraction is indicated in figure B. It yields a vector that is nearly perpendicular to $\boldsymbol{r}(t)$ and $\boldsymbol{r}(t + \Delta t)$. Indeed, the instantaneous velocity, found by allowing Δt to shrink to zero, is a vector $\boldsymbol{v}$ that is perpendicular to $\boldsymbol{r}$ at every instant and whose magnitude is,

$$|v| = r\frac{d\theta}{dt} = r\omega.$$

The relationship between $\boldsymbol{r}$ and $\boldsymbol{v}$ is shown in figure C. It means that the particle's instantaneous velocity is always tangent to the circle.

Notice that, just as the position vector $\boldsymbol{r}$ may be described in terms of the components x and y given by equations $x = r\cos\theta = r\cos\omega t$, and $y = r\sin\theta = r\sin\omega t$, the velocity vector $\boldsymbol{v}$ may be described in terms of its projections on the x and y axes, given by,

$$v_x = \frac{dx}{dt} = -r\omega \sin\omega t,$$

$$v_y = \frac{dy}{dt} = r\omega \cos\omega t.$$

Imagine a new coordinate system, in which a vector of length ωr extends from the origin and points at all times in the same direction as $\boldsymbol{v}$. This construction is shown in Figure. Each time the particle sweeps out a complete circle, this vector also sweeps out a complete circle. In fact, its point is executing uniform circular motion at the same angular frequency as the particle itself. Because vectors have magnitude and direction, but not position in space, the vector that has been constructed is the velocity $\boldsymbol{v}$. The velocity of the particle is itself undergoing uniform circular motion at angular frequency ω.

Although the speed of the particle is constant, the particle is nevertheless accelerated, because its velocity is constantly changing direction. The acceleration $\boldsymbol{a}$ is given by,

$$a = \frac{dv}{dt}$$

Since $\boldsymbol{v}$ is a vector of length $r\omega$ undergoing uniform circular motion, equations

$\bar{v} = r\frac{\Delta\theta}{\Delta t}$. and $\bar{v} = \frac{r(t+\Delta t) - r(t)}{\Delta t}$. may be repeated, as illustrated in figure E above, giving

$$\bar{a} = r\omega\frac{\Delta\theta}{\Delta t}$$

$$\bar{a} = \frac{v(t+\Delta t) - v(t)}{\Delta t}.$$

Thus, one may conclude that the instantaneous acceleration is always perpendicular to $\boldsymbol{v}$ and its magnitude is,

$$|a| = r\omega\frac{d\theta}{dt} = r\omega^2.$$

Since $\boldsymbol{v}$ is perpendicular to $\boldsymbol{r}$, and $\boldsymbol{a}$ is perpendicular to $\boldsymbol{v}$, the vector $\boldsymbol{a}$ is rotated 180° with respect to $\boldsymbol{r}$. In other words, the acceleration is parallel to $\boldsymbol{r}$ but in the opposite direction. The same conclusion may be reached by realizing that $\boldsymbol{a}$ has x and y components given by,

$$a = \frac{dv}{dt} = -r\omega \cos\omega t,$$

$$a_y = \frac{dv_y}{dt} = -r\omega^2 \sin\omega t.$$

Similar to equations $v_x = \frac{dx}{dt} = -r\omega \sin \omega t$ and $v_y = \frac{dy}{dt} = r\omega \cos \omega t$.

When equations $a_x = \frac{dv_x}{dt} = -r\omega^2 \cos \omega t$ and $a_y = \frac{dv_y}{dt} = -r\omega^2 \sin \omega t$.

are compared with equations $x = r\cos\theta = r\cos\omega t$, and $y = r\sin\theta = r\sin\omega t$, for x and y, it is clear that the components of $\boldsymbol{a}$ are just those of $\boldsymbol{r}$ multiplied by $-\omega^2$, so that $\boldsymbol{a} = -\omega^2\boldsymbol{r}$. This acceleration is called the centripetal acceleration, meaning that it is inward, pointing along the radius vector toward the centre of the circle. It is sometimes useful to express the centripetal acceleration in terms of the speed v. Using $v = \omega r$, one can write,

$$a = -\frac{v^2}{r}$$

Circular Orbits

The detailed behaviour of real orbits is the concern of celestial mechanics.

A body in uniform circular motion undergoes at all times a centripetal acceleration given by equation $a = -\frac{v^2}{r}$. According to Newton's second law, a force is required to produce this acceleration. In the case of an orbiting planet, the force is gravity. The situation is illustrated in figure. The gravitational attraction of the Sun is an inward (centripetal) force acting on Earth. This force produces the centripetal acceleration of the orbital motion.

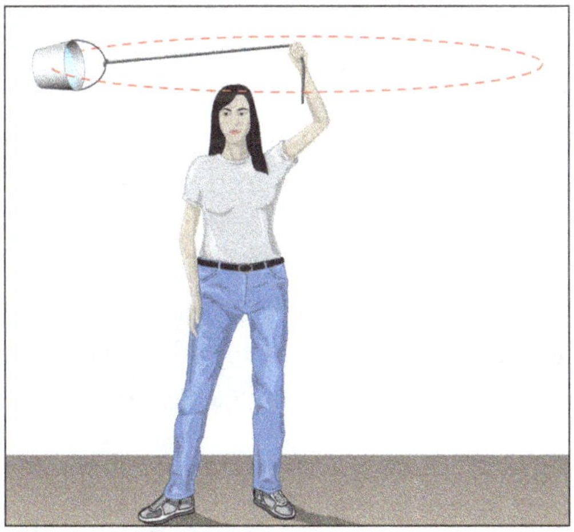

Centripetal force. The bucket experiences a centripetal force directed along the string toward the centre of the circle.

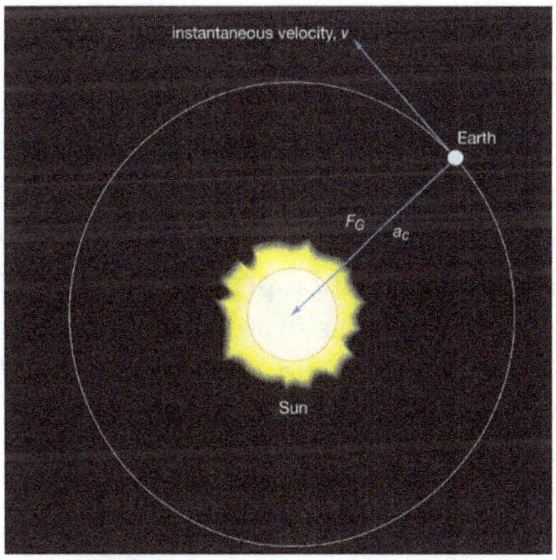

Figure: The gravitational force *FG* exerted by the Sun on the Earth produces the centripetal acceleration *ac* of the Earth's orbital motion.

Before these ideas are expressed quantitatively, an understanding of why a force is needed to maintain a body in an orbit of constant speed is useful. The reason is that, at each instant, the velocity of the planet is tangent to the orbit. In the absence of gravity, the planet would obey the law of inertia (Newton's first law) and fly off in a straight line in the direction of the velocity at constant speed. The force of gravity serves to overcome the inertial tendency of the planet, thereby keeping it in orbit.

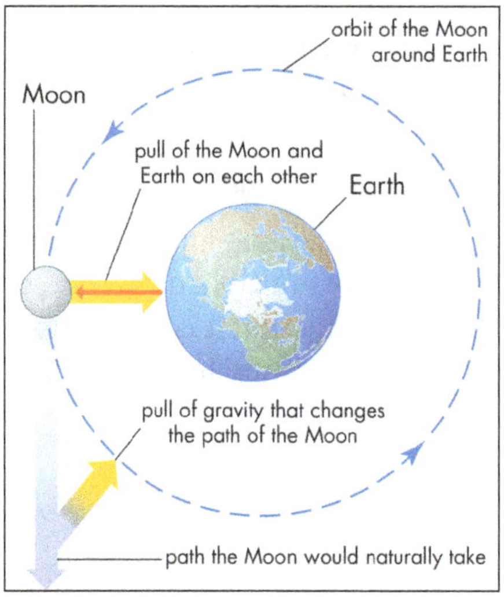

Effects of gravity on the Moon and Earth.
Effects of gravity on Earth and the Moon.

The gravitational force between two bodies such as the Sun and Earth is given by,

$$F = -G\frac{M_S M_E}{r^2},$$

where *MS* and *ME* are the masses of the Sun and Earth, respectively, *r* is the distance between their centres, and *G* is a universal constant equal to 6.674×10^{-11} Nm^2/kg^2 (Newton metres squared per kilogram squared). The force acts along the direction connecting the two bodies (i.e., along the radius vector of the uniform circular motion), and the minus sign signifies that the force is attractive, acting to pull Earth toward the Sun.

To an observer on the surface of Earth, the planet appears to be at rest at (approximately) a constant distance from the Sun. It would appear to the observer, therefore, that any force (such as the Sun's gravity) acting on Earth must be balanced by an equal and opposite force that keeps Earth in equilibrium. In other words, if gravity is trying to pull Earth into the Sun, some opposing force must be present to prevent that from happening. In reality, no such force exists. Earth is in freely accelerated motion caused by an unbalanced force. The apparent force, known in mechanics as a pseudoforce, is due to the fact that the observer is actually in accelerated motion. In the case of orbital motion, the outward pseudoforce that balances gravity is called the centrifugal force.

For a uniform circular orbit, gravity produces an inward acceleration given by equation, a = −v2/r. The pseudoforce f needed to balance this acceleration is just equal to the mass of Earth times an equal and opposite acceleration, or f = MEv2/r. The earthbound observer then believes that there is no net force acting on the planet—i.e., that F + f = 0, where F is the force of gravity given by equation $F = -G\frac{M_S M_E}{r^2}$. Combining these equations yields a relation between the speed v of a planet and its distance r from the Sun:

$$v^2 = G\frac{M_S}{r}$$

It should be noted that the speed does not depend on the mass of the planet. This occurs for exactly the same reason that all bodies fall toward Earth with the same acceleration and that the period of a pendulum is independent of its mass. An orbiting planet is in fact a freely falling body.

Equation $v^2 = G\frac{M_S}{r}$ is a special case (for circular orbits) of Kepler's third law, Using the fact that $v = 2\pi r/T$, where $2\pi r$ is the circumference of the orbit and T is the time to make a complete orbit (i.e., T is one year in the life of the planet), it is easy to show that $T^2 = (4\pi^2/GMS)r^3$. This relation also may be applied to satellites in circular orbit around Earth (in which case, *ME* must be substituted for *MS*) or in orbit around any other central body.

Angular Momentum and Torque

A particle of mass m and velocity $\boldsymbol{v}$ has linear momentum $\boldsymbol{p} = m\boldsymbol{v}$. The particle may also have angular momentum $\boldsymbol{L}$ with respect to a given point in space. If $\boldsymbol{r}$ is the vector from the point to the particle, then

$$L = r \times P$$

Notice that angular momentum is always a vector perpendicular to the plane defined by the vectors $\boldsymbol{r}$ and $\boldsymbol{p}$ (or $\boldsymbol{v}$). For example, if the particle (or a planet) is in a circular orbit, its angular momentum with respect to the centre of the circle is perpendicular to the plane of the orbit and in the direction given by the vector cross product right-hand rule, as shown in figure. Moreover, since in the case of a circular orbit, $\boldsymbol{r}$ is perpendicular to $\boldsymbol{p}$ (or $\boldsymbol{v}$), the magnitude of $\boldsymbol{L}$ is simply,

$$L = rP = mvr$$

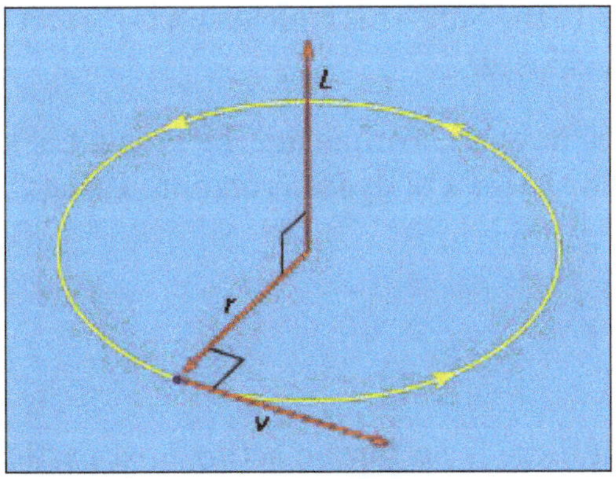

Figure: The angular momentum L of a particle traveling in a circular orbit.

The significance of angular momentum arises from its derivative with respect to time,

$$\frac{dL}{dt} = \frac{d}{dt}(r \times p) = m\frac{d}{dt}(r \times v),$$

where $\boldsymbol{p}$ has been replaced by $m\boldsymbol{v}$ and the constant m has been factored out. Using the product rule of differential calculus,

$$\frac{d}{dt}(r \times v) = \frac{dr}{dt} \times v + r \times \frac{dv}{dt}.$$

In the first term on the right-hand side of equation $\frac{d}{dt}(r \times v) = \frac{dr}{dt} \times v + r \times \frac{dv}{dt}$ ***dr***/*dt* is simply the velocity ***v***, leaving ***v*** × ***v***. Since the cross product of any vector with itself is always zero, that term drops out, leaving

$$\frac{d}{dt}(r \times v) = r \times \frac{dv}{dt}.$$

Here, ***dv***/*dt* is the acceleration *a* of the particle. Thus, if equation $\frac{d}{dt}(r \times v) = r \times \frac{dv}{dt}$ is multiplied by *m*, the left-hand side becomes ***dL***/*dt*, as in equation $\frac{dL}{dt} = \frac{d}{dt}(r \times p) = m\frac{d}{dt}(r \times v)$, and the right-hand side may be written ***r*** × ***ma***. Since, according to Newton's second law, ***ma*** is equal to ***F***, the net force acting on the particle, the result is

$$\frac{dL}{dt} r \times F.$$

Equation $\frac{dL}{dt} r \times F$ means that any change in the angular momentum of a particle must be produced by a force that is not acting along the same direction as ***r***. One particularly important application is the solar system. Each planet is held in its orbit by its gravitational attraction to the Sun, a force that acts along the vector from the Sun to the planet. Thus, the force of gravity cannot change the angular momentum of any planet with respect to the Sun. Therefore, each planet has constant angular momentum with respect to the Sun. This conclusion is correct even though the real orbits of the planets are not circles but ellipses.

The quantity ***r*** × ***F*** is called the torque τ. Torque may be thought of as a kind of twisting force, the kind needed to tighten a bolt or to set a body into rotation. Using this definition, equation above may be rewritten,

$$\tau = r \times F = \frac{dL}{dt}.$$

Equation above means that if there is no torque acting on a particle, its angular momentum is constant, or conserved. Suppose, however, that some agent applies a force ***Fa*** to the particle resulting in a torque equal to ***r*** × ***Fa***. According to Newton's third law, the particle must apply a force −***Fa*** to the agent. Thus, there is a torque equal to −***r*** × ***Fa*** acting on the agent. The torque on the particle causes its angular momentum to change at a rate given by ***dL***/*dt* = ***r*** × ***Fa***. However, the angular momentum ***La*** of the agent is changing at the rate ***dLa***/*dt* = −***r*** × ***Fa***. Therefore, ***dL***/*dt* + ***dLa***/*dt* = 0, meaning that the total angular momentum of particle plus agent is constant, or conserved. This principle may be generalized to include all interactions between bodies of any kind, acting by way of forces of any kind. Total angular momentum is always conserved. The law of conservation of angular momentum is one of the most important principles in all of physics.

Motion of a Group Of Particles

Centre of Mass

The word particle has been used in this topic to signify an object whose entire mass is concentrated at a point in space. In the real world, however, there are no particles of this kind. All real bodies have sizes and shapes. Furthermore, as Newton believed and is now known, all bodies are in fact compounded of smaller bodies called atoms. Therefore, the science of mechanics must deal not only with particles but also with more complex bodies that may be thought of as collections of particles.

To take a specific example, the orbit of a planet around the Sun was discussed earlier as if the planet and the Sun were each concentrated at a point in space. In reality, of course, each is a substantial body. However, because each is nearly spherical in shape, it turns out to be permissible, for the purposes of this problem, to treat each body as if its mass were concentrated at its centre. This is an example of an idea that is often useful in discussing bodies of all kinds: the centre of mass. The centre of mass of a uniform sphere is located at the centre of the sphere. For many purposes (such as the one cited above) the sphere may be treated as if all its mass were concentrated at its centre of mass.

To extend the idea farther, consider Earth and the Sun not as two separate bodies but as a single system of two bodies interacting with one another by means of the force of gravity. the Sun was assumed to be at rest at the centre of the orbit, but, according to Newton's third law, it must actually be accelerated by a force due to Earth that is equal and opposite to the force that the Sun exerts on Earth. In other words, considering only the Sun and Earth (ignoring, for example, all the other planets), if *MS* and *ME* are, respectively, the masses of the Sun and Earth, and if $\boldsymbol{a}_S$ and $\boldsymbol{a}_E$ are their respective accelerations, then combining Newton's second and third laws results in the equation $M_S\boldsymbol{a}_S = -M_E\boldsymbol{a}_E$. Writing each $\boldsymbol{a}$ as $d\boldsymbol{v}/dt$, this equation is easily manipulated to give,

$$\frac{d}{dt}\left(M_S v_S + M_E v_E\right) = 0,$$

$$M_S v_S + M_E v_E = \text{constant}.$$

This remarkable result means that, as Earth orbits the Sun and the Sun moves in response to Earth's gravitational attraction, the entire two-body system has constant linear momentum, moving in a straight line at constant speed. Without any loss of generality, one can imagine observing the system from a frame of reference moving along with that same speed and direction. This is sometimes called the centre-of-mass frame. In this frame, the momentum of the two-body system—i.e., the constant in equation $M_S v_S + M_E v_E = \text{constant}$ —is equal to zero. Writing each of the $\boldsymbol{v}$'s as the corresponding $d\boldsymbol{r}/dt$, equation $M_S v_S + M_E v_E = \text{constant}$ may be expressed in the form,

$$\frac{d}{dt}\left(M_S r_S + M_E r_E\right) = 0.$$

Thus, $M_S\boldsymbol{r}_S$ and $M_E\boldsymbol{r}_E$ are two vectors whose vector sum does not change with time. The sum is defined to be the constant vector $\boldsymbol{MR}$, where M is the total mass of the system and equals $M_S + M_E$. Thus,

$$MR = M_S r_S + M_E r_E.$$

This procedure defines a constant vector $\boldsymbol{R}$, from any arbitrarily chosen point in space. The relation between vectors $\boldsymbol{R}$, $\boldsymbol{r}_S$, and $\boldsymbol{r}_E$ is shown in Figure. The fact that $\boldsymbol{R}$ is constant (although $\boldsymbol{r}_S$ and $\boldsymbol{r}_E$ are not constant) means that, rather than Earth orbiting the Sun, Earth and the Sun are both orbiting an imaginary point fixed in space. This point is known as the centre of mass of the two-body system.

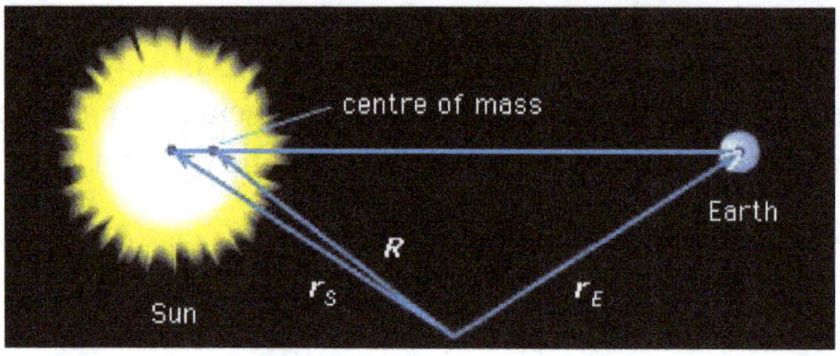

Figure: The centre of mass of the two-body Earth-Sun system.

Knowing the masses of the two bodies ($M_S = 1.99 \times 10^{30}$ kilograms, $M_E = 5.98 \times 10^{24}$ kilograms), it is easy to find the position of the centre of mass. The origin of the coordinate system may be chosen to be located at the centre of mass merely by defining $\boldsymbol{R} = 0$. Then $r_S = (M_E/M_S)\, r_E \approx 450$ kilometres, when r_E is rounded to 1.5×10^8 km. A few hundred kilometres is so small compared to r_E that, for all practical purposes, no appreciable error occurs when r_S is ignored and the Sun is assumed to be stationary at the centre of the orbit.

With this example as a guide, it is now possible to define the centre of mass of any collection of bodies. Assume that there are N bodies altogether, each labeled with numbers ranging from 1 to N, and that the vector from an arbitrary origin to the ith body—where i is some number between 1 and N—is $\boldsymbol{r}_i$, as shown in figure. Let the mass of the ith body be m_i. Then the total mass of the N-body system is,

$$m = \sum_{i=1}^{N} m,$$

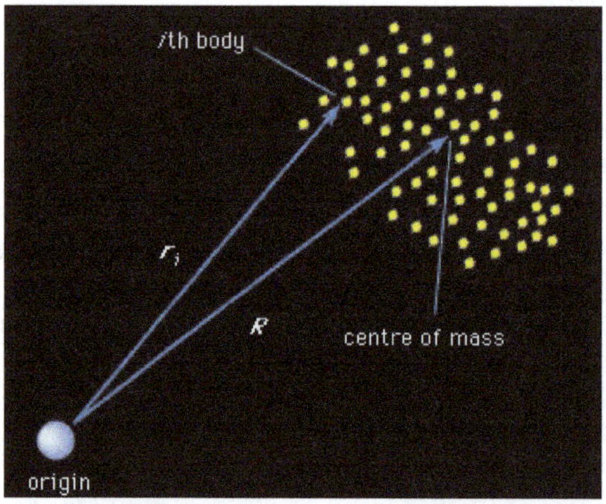

Figure: The centre of mass of an N-body system.

and the centre of mass of the system is found at the end of a vector $\boldsymbol{R}$ given by,

$$mR = \sum_{i=1}^{N} m_i r_i ,$$

as illustrated in figure. This definition applies regardless of whether the N bodies making up the system are the stars in a galaxy, the atoms in a rigid body, larger and arbitrarily chosen segments of a rigid body, or any other system of masses. According to equation $mR = \sum_{i=1}^{N} m_i r_i$, the vector to the centre of mass of any system is a kind of weighted average of the vectors to all the components of the system.

The statics and dynamics of many complicated bodies or systems may often be understood by simply applying Newton's laws as if the system's mass were concentrated at the centre of mass.

Conservation of Momentum

Newton's second law, in its most general form, says that the rate of a change of a particle's momentum $\boldsymbol{p}$ is given by the force acting on the particle; i.e., $\boldsymbol{F} = d\boldsymbol{p}/dt$. If there is no force acting on the particle, then, since $d\boldsymbol{p}/dt = 0$, $\boldsymbol{p}$ must be constant, or conserved. This observation is merely a restatement of Newton's first law, the principle of inertia: if there is no force acting on a body, it moves at constant speed in a straight line.

Now suppose that an external agent applies a force $\boldsymbol{F}_a$ to the particle so that $\boldsymbol{p}$ changes according to,

$$\frac{dp}{dt} = F_a.$$

According to Newton's third law, the particle must apply an equal and opposite force $-\boldsymbol{F}_a$ to the external agent. The momentum $\boldsymbol{p}_a$ of the external agent therefore changes according to,

$$\frac{dp_a}{dt} = F_a.$$

Adding together equations $\frac{dp}{dt} = F_a.$ and $\frac{dp_a}{dt} = F_a.$ results in the equation,

$$\frac{dp}{dt}(p + p_a) = 0.$$

The force applied by the external agent changes the momentum of the particle, but at the same time the momentum of the external agent must also change in such a way that the total momentum of both together is constant, or conserved. This idea may be generalized to give the law of conservation of momentum: in all the interactions between all the bodies in the universe, total momentum is always conserved.

It is useful in this light to examine the behaviour of a complicated system of many parts. The centre of mass of the system may be found using equation $mR = \sum_{i=1}^{N} m_i r_i$. Differentiating with respect to time gives,

$$mv = \sum_{i=1}^{N} m_i v_i,$$

where $\boldsymbol{v} = d\boldsymbol{R}/dt$ and $\boldsymbol{vi} = d\boldsymbol{ri}/dt$. Note that $\boldsymbol{mivi}$ is the momentum of the ith part of the system, and $m\boldsymbol{v}$ is the momentum that the system would have if all its mass (i.e., m) were concentrated at its centre of mass, the point whose velocity is $\boldsymbol{v}$. Thus, the momentum associated with the centre of mass is the sum of the momenta of the parts.

Suppose now that there is no external agent applying a force to the entire system. Then the only forces acting on the system are those exerted by the parts on one another. These forces may accelerate the individual parts. Differentiating equation $mv = \sum_{i=1}^{N} m_i v_i$, with respect to time gives,

$$m\frac{dv}{dt} = \sum_{i=1}^{N} m_i \frac{dv_i}{dt} = \sum_{i=1}^{N} F_i$$

where $\boldsymbol{F}_i$ is the net force, or the sum of the forces, exerted by all the other parts of the body on the ith part. $\boldsymbol{F}_i$ is defined mathematically by the equation,

$$F_i = \sum_{j=1}^{N} F_{ij},$$

where $\boldsymbol{F}_{ij}$ represents the force on body i due to body j (the force on body i due to itself, $\boldsymbol{F}_{ii}$, is zero). The motion of the centre of mass is then given by the complicated-looking formula,

$$m\frac{dv}{dt} = \sum_{i=1}^{N}\left(\sum_{j=1}^{N} F_{ij}\right),$$

This complicated formula may be greatly simplified, however, by noting that Newton's third law requires that for every force $\boldsymbol{F}_{ij}$ exerted by the jth body on the ith body, there is an equal and opposite force $-\boldsymbol{F}_{ij}$ exerted by the ith body on the jth body. In other words, every term in the double sum has an equal and opposite term. The double summation on the right-hand side of equation $F_i = \sum_{j=1}^{N} F_{ij}$, always adds up to zero. This result is true regardless of the complexity of the system, the nature of the forces acting between the parts, or the motions of the parts. In short, in the absence of external forces acting on the system as a whole, $m d\boldsymbol{v}/dt = 0$, which means that the momentum of the centre of mass of the system is always conserved. Having determined that momentum is conserved whether or not there is an external force acting, one may conclude that the total momentum of the universe is always conserved.

Collisions

A collision is an encounter between two bodies that alters at least one of their courses. Altering the course of a body requires that a force be applied to it. Thus, each body exerts a force on the other. These forces of interaction may operate at some distance, as do the gravitational and electromagnetic forces, or the bodies may appear to make physical contact. However, even apparent contact between two bodies is only a macroscopic manifestation of microscopic forces that act between atoms some distance apart. There is no fundamental distinction between physical contact and interaction at a distance.

The importance of understanding the mechanics of collisions is obvious to anyone who has ever driven an automobile. In modern physics, however, collisions are important for a different reason. The current understanding of the subatomic particles of which atoms are composed is derived entirely from studying the results of collisions among them. Thus, in modern physics, the description of collisions is a significant part of the understanding of matter. These descriptions are quantum mechanical rather than classical, but they are nevertheless closely based on principles that arise out of classical mechanics.

It is possible in principle to predict the result of a collision using Newton's second law directly. Suppose that two bodies are going to collide and that $\boldsymbol{F}$, the force of interaction between them, is known to be a function of $\boldsymbol{r}$, the distance between them. Then, if it is known that, say, one particle has incident momentum $\boldsymbol{p}$, the problem is solved if the final momentum $\boldsymbol{p} + \Delta\boldsymbol{p}$ can be determined. Inverting Newton's second law, $\boldsymbol{F} = d\boldsymbol{p}/dt$, the change in momentum is given by,

$$\Delta p = \int_{-\infty}^{\infty} F dt.$$

This integral is known as the impulse imparted to the particle. In order to perform the integral, it is necessary to know $\boldsymbol{r}$ at all times so that $\boldsymbol{F}$ may be known at all times. More realistically, $\Delta\boldsymbol{p}$ is the sum of a series of small steps, such that

$$\delta p = F\delta t.$$

where $\boldsymbol{F}$ depends on the instantaneous distance between the particles. Because $\boldsymbol{p} = m\boldsymbol{v} = md\boldsymbol{r}/dt$, the change in $\boldsymbol{r}$ in this step is,

$$\delta r = \frac{p}{m}\delta t.$$

At the next step, there is a new distance, $\boldsymbol{r} + \delta\boldsymbol{r}$, giving a new value of the force in equation $\delta p = F\delta t.$ and a new momentum, $\boldsymbol{p} + \delta\boldsymbol{p}$, in equation $\delta r = \frac{p}{m}\delta t.$ This method of analyzing collisions is used in numerical calculations on digital computers.

To predict the result of a collision analytically (rather than numerically) it is often most useful to apply conservation laws. In any collision (as in any other phenomenon), energy, momentum, and angular momentum are always conserved. Judicious application of these laws may be extremely useful because they do not depend in any way on the detailed nature of the interaction (i.e., the force as a function of distance).

This point can be illustrated by the following example. A collision is to take place between two bodies of the same mass m. One of the bodies is initially at rest (its momentum is zero). The other has initial momentum $\boldsymbol{p}_0$. After the collision, the body previously at rest has momentum $\boldsymbol{p}_1$, and the body initially in motion has momentum $\boldsymbol{p}_2$. Since momentum is conserved, the total momentum after the collision, $\boldsymbol{p}_1 + \boldsymbol{p}_2$, must be equal to the total momentum before the collision, $\boldsymbol{p}_0$; that is,

$$p_0 = p_1 + p_2,$$

Equation $p_0 = p_1 + p_2,$ is the equation of a vector triangle, as shown in Figure. However, $\boldsymbol{p}_1$ and $\boldsymbol{p}_2$ are not determined by this condition; they are only constrained by it.

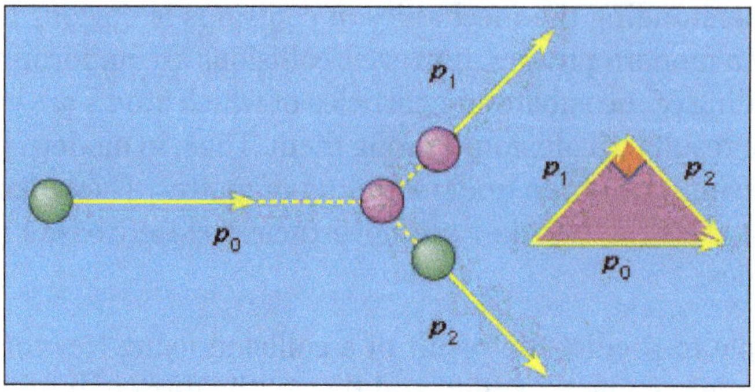

Figure: Collision between two particles of equal mass.

Although energy is always conserved, the kinetic energy of the incident body is not always converted entirely into the kinetic energy of the two bodies after the collision. For example, if the bodies are microscopic (say, two identical atoms), the collision may cause one or both to be excited into a state of higher internal energy than it started with. Such an event would leave correspondingly less kinetic energy for the outgoing atoms. In fact, it is precisely by studying the trajectories of outgoing projectiles in collisions like these that physicists are able to determine the possible excited states of microscopic particles.

In a collision between macroscopic objects, some of the kinetic energy is always converted to heat. Heat is the energy of random vibrations of the atoms and molecules that constitute the bodies. However, if the amount of heat is negligible compared to the initial kinetic energy, it may be ignored. Such a collision is said to be elastic.

Suppose the collision between two bodies, each of mass m, is between billiard balls, and suppose it is elastic (a reasonably good approximation of real billiard balls). The kinetic energy of the incident ball is then equal to the sum of the kinetic energies of the outgoing balls. According to equation $K = \frac{1}{2}mv^2$. the kinetic energy of a moving object is given by $K = {}^1/_2 mv^2$, where v is the speed of the ball (technically, the energy associated with the fact that the ball is rolling as well as translating is ignored here;

Equation $K = \frac{1}{2}mv^2$. may be written in a particularly useful form by recognizing that since $p = mv$,

$$K = \frac{1}{2}mv^2 = \frac{p^2}{2m}.$$

Then the conservation of kinetic energy may be written,

$$\frac{p_0^2}{2m} = \frac{p_1^2}{2m} + \frac{p_2^2}{2m}$$

or, canceling the factors $2m$,

$$p_0^2 = p_1^2 + p_2^2$$

Comparing this result with equation $p_0 = p_1 + p_2$, shows that the vector triangle is pythagorean; $\boldsymbol{p}_1$ and $\boldsymbol{p}_2$ are perpendicular. This result is well known to all experienced pool players. Notice that it was possible to arrive at this result without any knowledge of the forces that act when billiard balls collide.

Relative Motion

A collision between two bodies can always be described in a frame of reference in which the total momentum is zero. This is the centre-of-mass (or centre-of-momentum) frame. Then, for example, in the collision between two bodies of the same mass the two bodies always have equal and opposite velocities, as shown in figure. It should be noted that, in this frame of reference, the outgoing momenta are antiparallel and not perpendicular.

Any collection of bodies may similarly be described in a frame of reference in which the total momentum is zero. This frame is simply the one in which the centre of mass is at rest. This fact is easily seen by differentiating equation $mR = \sum_{i=1}^{N} m_i r_i$, with respect to time, giving,

$$m\frac{dR}{dt} = \sum_{i=1}^{N} m_i \frac{dr_i}{dt}.$$

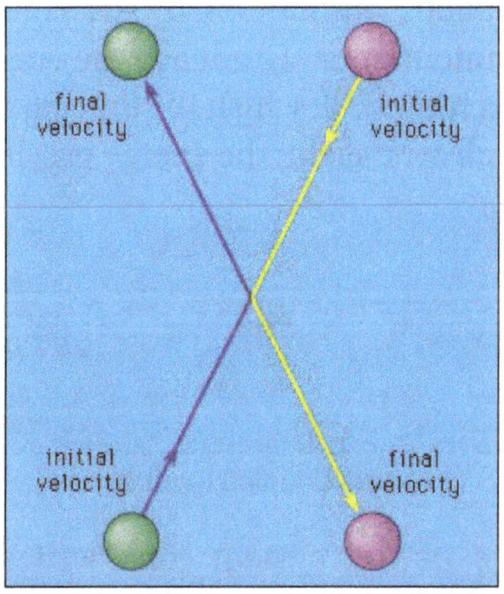

Figure: Collision between two particles of equal mass as seen from the centre-of-mass frame of reference.

The right-hand side is the sum of the momenta of all the bodies. It is equal to zero if the velocity of the centre of mass, $d\mathbf{R}/dt$, is equal to zero.

If Newton's second law is correct in any frame of reference, it will also appear to be correct to an observer moving with any constant velocity with respect to that frame. This principle, called the principle of Galilean relativity, is true because, to the moving observer, the same constant velocity seems to have been added to the velocity of every particle in the system. This change does not affect the accelerations of the particles (since the added velocity is constant, not accelerated) and therefore does not change the apparent force (mass times acceleration) acting on each particle. That is why it is permissible to describe a problem from the centre-of-momentum frame (provided that the centre of mass is not accelerated) or from any other frame moving at constant velocity with respect to it.

If this principle is strictly correct, the fundamental forces of physics should not contain any particular speed. This must be true because the speed of any object will be different to observers in different but equally good frames of reference, but the force should always be the same. It turns out, according to the theory of James Clerk Maxwell, that there is an intrinsic speed in the force laws of electricity and magnetism: the speed of light appears in the forces between electric charges and between magnetic poles. This discrepancy was ultimately resolved by Albert Einstein's special theory of relativity. According to the special theory of relativity, Newtonian mechanics breaks down when the relative speed between particles approaches the speed of light.

Coupled Oscillators

The motion of a group of particles bound by springs to one another is discussed. The solutions of this seemingly academic problem have far-reaching implications in many fields of physics. For example, a system of particles held together by springs turns out to be a useful model of the behaviour of atoms mutually bound in a crystalline solid.

To begin with a simple case, consider two particles in a line, as shown in figure. Each particle has mass m, each spring has spring constant k, and motion is restricted to the horizontal, or x, direction. Even this elementary system is capable of surprising behaviour, however. For instance, if one particle is held in place while the other is displaced, and then both are released, the displaced particle immediately begins to execute simple harmonic motion. This motion, by stretching the spring between the particles, starts to excite the second particle into motion. Gradually the energy of motion passes from the first particle to the second until a point is reached at which the first particle is at rest and only the second is oscillating. Then the process starts all over again, the energy passing in the opposite direction.

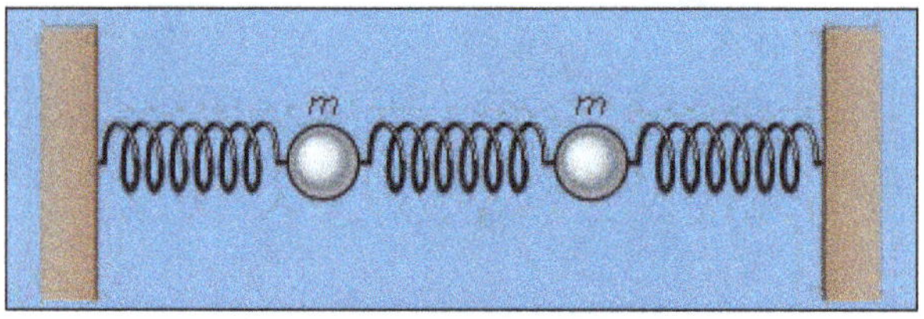

Figure: Coupled oscillators.

To analyze the possible motions of the system, one writes equations similar to equation $a = \frac{d^2x}{dt^2} = -\frac{k}{m}x$, giving the acceleration of each particle owing to the forces acting on it. There is one equation for each particle (two equations in this case). The force on each particle depends not only on its displacement from its equilibrium position but also on its distance from the other particle, since the spring between them stretches or compresses according to that distance. For this reason the motions are coupled, the solution of each equation (the motion of each particle) depending on the solution of the other (the motion of the other).

Analyzing the system yields the fact that there are two special states of motion in which both particles are always in oscillation with the same frequency. In one state, the two particles oscillate in opposite directions with equal and opposite displacements from equilibrium at all times. In the other state, both particles move together, so that the spring between them is never stretched or compressed. The first of these motions has higher frequency than the second because the centre spring contributes an increase in the restoring force.

These two collective motions, at different, definite frequencies, are known as the normal modes of the system.

If a third particle is inserted into the system together with another spring, there will be three equations to solve, and the result will be three normal modes. A large number N of particles in a line will have N normal modes. Each normal mode has a definite frequency at which all the particles oscillate. In the highest frequency mode each particle moves in the direction opposite to both of its neighbours. In the lowest frequency mode, neighbours move almost together, barely disturbing the springs between them. Starting from one end, the amplitude of the motion gradually builds up, each particle moving a bit more than the one before, reaching a maximum at the centre, and then decreasing again. A plot of the amplitudes, shown in figure, basically describes one-half of a sine wave from one end of the system to the other. The next mode is a full sine wave, then $^3/_2$ of a sine wave, and so on to the highest frequency mode, which may be visualized as $N/2$ sine waves. If the vibrations were up and down rather than side to side, these modes would be identical to the fundamental and harmonic vibrations excited by plucking a guitar string.

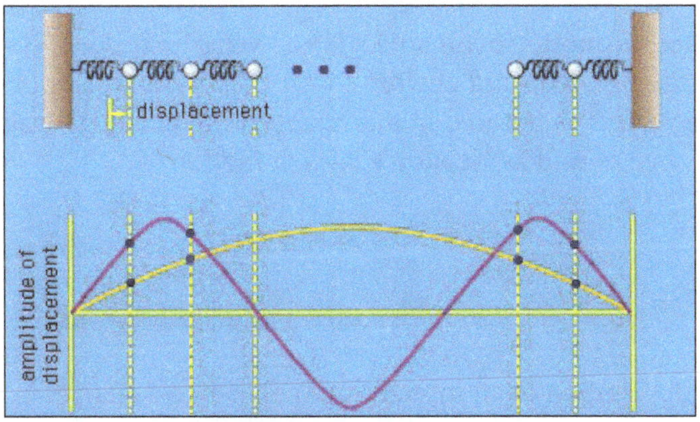

Figure: Normal modes.

The atoms of a crystal are held in place by mutual forces of interaction that oppose any disturbance from equilibrium positions, just as the spring forces in the example above. For small displacements of the atoms, they behave mathematically just like spring forces—i.e., they obey Hooke's law, equation $F = -kx$. Each atom is free to move in three dimensions rather than one, however; therefore each atom added to a crystal adds three normal modes. In a typical crystal at ordinary temperature, all these modes are always excited by random thermal energy. The lower-frequency, longer-wavelength modes may also be excited mechanically. These are called sound waves.

References

- Kinematics, science: britannica.com, Retrieved 31 March, 2019
- Distance-and-displacement, physics: vedantu.com, Retrieved 14 July, 2019
- stewart, james (2001). "§2.8 - the derivative as a function". Calculus (2nd ed.). Brooks/cole. Isbn 0-534-37718-1
- Distance-and-displacement, physics: byjus.com, Retrieved 17 May, 2019
- iec 60050 - details for iev number 113-01-33: "speed"". Electropedia: the world's online electrotechnical vocabulary. Retrieved 2017-06-08.
- Robert resnick and jearl walker, fundamentals of physics, wiley; 7 sub edition (june 16, 2004). Isbn 0-471-23231-9.
- Motion-mechanics, science: britannica.com, Retrieved 19 April, 2019
- Motion-of-a-particle-in-one-dimension, mechanics, science: britannica.com, Retrieved 5 February, 2019

3

Force and Newton's Laws of Motion

The action that changes the motion of an object is known as force. Conservative force, centripetal force , centrifugal force, inertia and newton's laws of motion are few of the concepts that fall within its domain. The topics elaborated in this chapter will help in gaining a better perspective about these types of force as well as Newton's laws of motion.

Force

Force is a quantitative description of an interaction that causes a change in an object's motion. An object may speed up, slow down, or change direction in response to a force. Put another way, force is any action that tends to maintain or alter the motion of a body or to distort it. Objects are pushed or pulled by forces acting on them.

Contact force is defined as the force exerted when two physical objects come in direct contact with each other. Other forces, such as gravitation and electromagnetic forces, can exert themselves even across the empty vacuum of space.

Units of Force

Force is a vector; it has both direction and magnitude. The SI unit for force is the newton (N). One newton of force is equal to 1 kg * m/s^2 (where the "*" symbol stands for "times").

Force is proportional to acceleration, which is defined as the rate of change of velocity. In calculus terms, force is the derivative of momentum with respect to time.

Contact vs. Noncontact Force

There are two types of forces in the universe: contact and noncontact. Contact forces, as the name implies, take place when objects touch each other, such as kicking a ball: One object (your foot) touches the other object (the ball). Noncontact forces are those where objects do not touch each other.

Contact forces can be classified according to six different types:

- Tensional: such as a string being pulled tight.

- Spring: such as the force exerted when you compress two ends of a spring.
- Normal reaction: where one body provides a reaction to a force exerted upon it, such as a ball bouncing on a blacktop.
- Friction: the force exerted when an object moves across another, such as a ball rolling over a blacktop.
- Air friction: the friction that occurs when an object, such as a ball, moves through the air.
- Weight: where a body is pulled toward the center of the Earth due to gravity.

Noncontact forces can be classified according to three types:

- Gravitational: which is due to the gravitational attraction between two bodies.
- Electrical: which is due to the electrical charges present in two bodies.
- Magnetic: which occurs due to the magnetic properties of two bodies, such as the opposite poles of two magnets being attracted to each other.

Fundamental Forces

There are four fundamental forces that govern the interactions of physical systems. Scientists continue to pursue a unified theory of these forces:

- Gravitation: The force that acts between masses. All particles experience the force of gravity. If you hold a ball up in the air, for example, the mass of the Earth allows the ball to fall due to the force of gravity. Or if a baby bird crawls out of its nest, the gravity from the Earth will pull it to the ground. While the graviton has been proposed as the particle mediating gravity, it has not yet been observed.
- Electromagnetic: The force that acts between electrical charges. The mediating particle is the photon. For example, a loudspeaker uses the electromagnetic force to propagate the sound, and a bank's door locking system uses electromagnetic forces to help shut the vault doors tightly. Power circuits in medical instruments like magnetic resonance imaging use electromagnetic forces, as do the magnetic rapid transit systems in Japan and China—called "maglev" for magnetic levitation.
- Strong nuclear: The force that holds the nucleus of the atom together, mediated by gluons acting on quarks, antiquarks, and the gluons themselves. (A gluon is a messenger particle that binds quarks within the protons and neutrons. Quarks are fundamental particles that combine to form protons and neutrons, while antiquarks are identical to quarks in mass but opposite in electric and magnetic properties.)
- Weak nuclear: The force that is mediated by exchanging W and Z bosons and is seen in beta decay of neutrons in the nucleus. (A boson is a type of particle that obeys the rules of Bose-Einstein statistics.) At very high temperatures, the weak force and the electromagnetic force are indistinguishable.

Conservative Force

A conservative force is a force with the property that the total work done in moving a particle between two points is independent of the taken path. Equivalently, if a particle travels in a closed loop, the total work done (the sum of the force acting along the path multiplied by the displacement) by a conservative force is zero.

A conservative force depends only on the position of the object. If a force is conservative, it is possible to assign a numerical value for the potential at any point, and conversely. When an object moves from one location to another, the force changes the potential energy of the object by an amount that does not depend on the path taken, contributing to the mechanical energy and the overall conservation of energy. If the force is not conservative, then defining a scalar potential is not possible, because taking different paths would lead to conflicting potential differences between the start and end points.

Gravitational force is an example of a conservative force, while frictional force is an example of a non-conservative force.

Other examples of conservative forces are: force in elastic spring, electrostatic force between two electric charges, and magnetic force between two magnetic poles. The last two forces are called central forces as they act along the line joining the centres of two charged/magnetized bodies. A central force is conservative if it is spherically symmetric.

Informally, a conservative force can be thought of as a force that *conserves* mechanical energy. Suppose a particle starts at point A, and there is a force F acting on it. Then the particle is moved around by other forces, and eventually ends up at A again. Though the particle may still be moving, at that instant when it passes point A again, it has traveled a closed path. If the net work done by F at this point is 0, then F passes the closed path test. Any force that passes the closed path test for all possible closed paths is classified as a conservative force.

The gravitational force, spring force, magnetic force and electric force (at least in a time-independent magnetic field) are examples of conservative forces, while friction and air drag are classical examples of non-conservative forces.

For non-conservative forces, the mechanical energy that is lost (not conserved) has to go somewhere else, by conservation of energy. Usually the energy is turned into heat, for example the heat generated by friction. In addition to heat, friction also often produces some sound energy. The water drag on a moving boat converts the boat's mechanical energy into not only heat and sound energy, but also wave energy at the edges of its wake. These and other energy losses are irreversible because of the second law of thermodynamics.

Path Independence

A direct consequence of the closed path test is that the work done by a conservative force on a particle moving between any two points does not depend on the path taken by the particle.

This is illustrated in the figure below: The work done by the gravitational force on an object depends only on its change in height because the gravitational force is conservative. The work done by a conservative force is equal to the negative of change in potential energy during that process. For a proof, imagine two paths 1 and 2, both going from point A to point B. The variation of energy for the particle, taking path 1 from A to B and then path 2 backwards from B to A, is 0; thus, the work is the same in path 1 and 2, i.e., the work is independent of the path followed, as long as it goes from A to B.

For example, if a child slides down a frictionless slide, the work done by the gravitational force on the child from the start of the slide to the end is independent of the shape of the slide; it only depends on the vertical displacement of the child.

Mathematical Description

A force field *F*, defined everywhere in space (or within a simply-connected volume of space), is called a *conservative force* or *conservative vector field* if it meets any of these three *equivalent* conditions:

- The curl of *F* is the zero vector:

  $$\nabla \times \vec{F} = \vec{0}.$$

- There is zero net work (*W*) done by the force when moving a particle through a trajectory that starts and ends in the same place:

  $$W \equiv \oint_C \vec{F} \cdot \mathrm{d}\vec{r} = 0.$$

- The force can be written as the negative gradient of a potential, Φ:

  $$\vec{F} = -\nabla \Phi.$$

Proof that these three conditions are equivalent when f is a force field:

- 1 implies 2:

  Let *C* be any simple closed path (i.e., a path that starts and ends at the same point and has no self-intersections), and consider a surface *S* of which *C* is the boundary. Then Stokes' theorem says that,

  $$\int_S \left(\nabla \times \vec{F}\right) \cdot \mathrm{d}\vec{a} = \oint_C \vec{F} \cdot \mathrm{d}\vec{r}$$

  If the curl of F is zero the left hand side is zero - therefore statement 2 is true.

- 2 implies 3:

  Assume that statement 2 holds. Let *c* be a simple curve from the origin to a point x and define a function,

  $$\Phi(x) = -\int F \cdot \mathrm{d}r.$$

  The fact that this function is well-defined (independent of the choice of *c*) follows from statement 2. Anyway, from the fundamental theorem of calculus, it follows that,

  $$\vec{F} = -\nabla \Phi.$$

So statement 2 implies statement 3.

- 3 implies 1:

  Finally, assume that the third statement is true. A well-known vector calculus identity states that the curl of the gradient of any function is 0. Therefore, if the third statement is true, then the first statement must be true as well. This shows that statement 1 implies 2, 2 implies 3, and 3 implies 1. Therefore, all three are equivalent, Q.E.D.. (The equivalence of 1 and 3 is also known as (one aspect of) Helmholtz's theorem

The term conservative force comes from the fact that when a conservative force exists, it conserves mechanical energy. The most familiar conservative forces are gravity, the electric force (in a time-independent magnetic field), and spring force.

Many forces (particularly those that depend on velocity) are not force fields. In these cases, the above three conditions are not mathematically equivalent. For example, the magnetic force satisfies condition 2 (since the work done by a magnetic field on a charged particle is always zero), but does not satisfy condition 3, and condition 1 is not even defined (the force is not a vector field, so one cannot evaluate its curl). Accordingly, some authors classify the magnetic force as conservative, while others do not. The magnetic force is an unusual case; most velocity-dependent forces, such as friction, do not satisfy any of the three conditions, and therefore are unambiguously non-conservative.

Non-conservative Force

Examples of nonconservative forces are friction and non-elastic material stress. Despite conservation of total energy, non-conservative forces can arise in classical physics due to neglected degrees of freedom or from time-dependent potentials. For instance, friction may be treated without violating conservation of energy by considering the motion of individual molecules; however that means every molecule's motion must be considered rather than handling it through statistical methods. For macroscopic systems the nonconservative approximation is far easier to deal with than millions of degrees of freedom.

General relativity is non-conservative, as seen in the anomalous precession of Mercury's orbit. However, general relativity does conserve a stress–energy–momentum pseudotensor.

Centripetal Force

A centripetal force is a force that makes a body follow a curved path. Its direction is always orthogonal to the motion of the body and towards the fixed point of the instantaneous center of curvature of the path. Isaac Newton described it as "a force by which bodies are drawn or impelled, or in any way tend, towards a point as to a centre". In Newtonian mechanics, gravity provides the centripetal force responsible for astronomical orbits.

One common example involving centripetal force is the case in which a body moves with uniform speed along a circular path. The centripetal force is directed at right angles to the motion and also along the radius towards the centre of the circular path. The mathematical description was derived in 1659 by the Dutch physicist Christiaan Huygens.

Formula

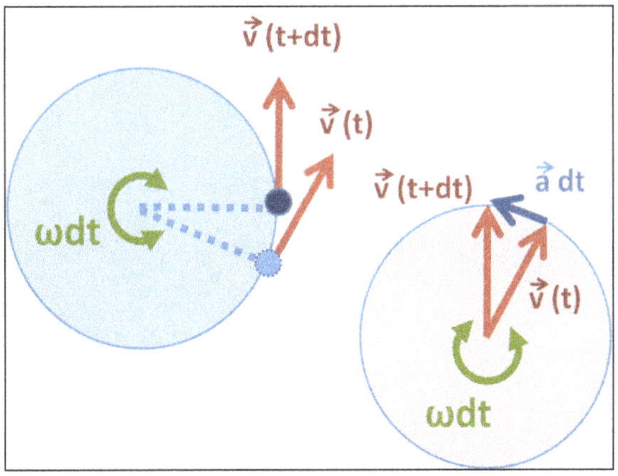

The magnitude of the centripetal force on an object of mass *m* moving at tangential speed *v* along a path with radius of curvature *r* is:

$$F_c = ma_c = \frac{mv^2}{r}$$

$$a_c = \frac{v}{t}\hat{r} = \frac{r\omega}{t}\hat{r} = v\omega = \frac{v^2}{r}$$

where a_c is the centripetal acceleration. The direction of the force is toward the center of the circle in which the object is moving, or the osculating circle (the circle that best fits the local path of the object, if the path is not circular). The speed in the formula is squared, so twice the speed needs four times the force. The inverse relationship with the radius of curvature shows that half the radial distance requires twice the force. This force is also sometimes written in terms of the angular velocity ω of the object about the center of the circle, related to the tangential velocity by the formula,

$$v = \omega r$$

so that

$$F_c = mr\omega^2.$$

Expressed using the orbital period *T* for one revolution of the circle,

$$\omega = \frac{2\pi}{T}$$

the equation becomes,

$$F_c = mr\left(\frac{2\pi}{T}\right)^2.$$

In particle accelerators, velocity can be very high (close to the speed of light in vacuum) so the same rest mass now exerts greater inertia (relativistic mass) thereby requiring greater force for the same centripetal acceleration, so the equation becomes:

$$F_c = \frac{\gamma m v^2}{r}$$

where

$$\gamma = \frac{1}{\sqrt{1 - \frac{v^2}{c^2}}}$$

is called the Lorentz factor.

More intuitively:

$$F_c = \gamma m v \omega$$

which is the rate of change of relativistic momentum ($\gamma m v$).

Sources

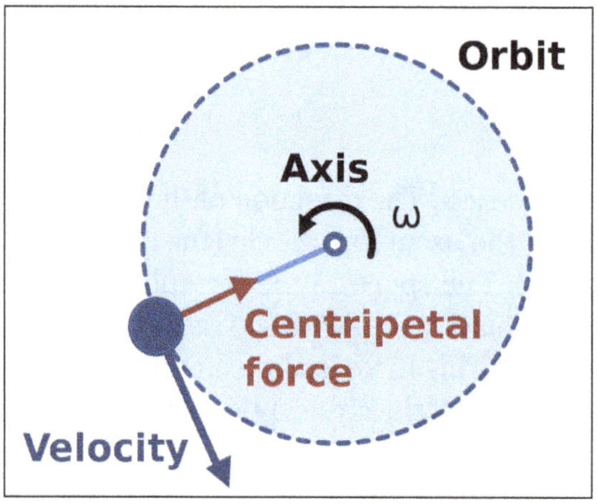

A body experiencing uniform circular motion requires a centripetal force, towards the axis as shown, to maintain its circular path.

In the case of an object that is swinging around on the end of a rope in a horizontal plane, the centripetal force on the object is supplied by the tension of the rope. The rope example is an example involving a 'pull' force. The centripetal force can also be supplied as a 'push' force, such as in the case where the normal reaction of a wall supplies the centripetal force for a wall of death rider.

Newton's idea of a centripetal force corresponds to what is nowadays referred to as a central force. When a satellite is in orbit around a planet, gravity is considered to be a centripetal force even though in the case of eccentric orbits, the gravitational force is directed towards the focus, and not towards the instantaneous center of curvature.

Another example of centripetal force arises in the helix that is traced out when a charged particle moves in a uniform magnetic field in the absence of other external forces. In this case, the magnetic force is the centripetal force that acts towards the helix axis.

Analysis of Several Cases

Below are three examples of increasing complexity, with derivations of the formulas governing velocity and acceleration.

Uniform Circular Motion

Uniform circular motion refers to the case of constant rate of rotation. Here are two approaches to describing this case.

Calculus Derivation

In two dimensions, the position vector $\mathbf{r}$, which has magnitude (length) r and directed at an angle θ above the x-axis, can be expressed in Cartesian coordinates using the unit vectors $\hat{x}$ and $\hat{y}$:

$$\mathbf{r} = r\cos(\theta)\hat{x} + r\sin(\theta)\hat{y}.$$

Assume uniform circular motion, which requires three things.

1. The object moves only on a circle.
2. The radius of the circle r does not change in time.
3. The object moves with constant angular velocity ω around the circle. Therefore, $\theta = \omega t$ where t is time.

Now find the velocity $\mathbf{v}$ and acceleration $\mathbf{a}$ of the motion by taking derivatives of position with respect to time.

$$\mathbf{r} = r\cos(\omega t)\hat{x} + r\sin(\omega t)\hat{y}$$

$$\dot{\mathbf{r}} = \mathbf{v} = -r\omega\sin(\omega t)\hat{x} + r\omega\cos(\omega t)\hat{y}$$

$$\ddot{\mathbf{r}} = \mathbf{a} = -r\omega^2\cos(\omega t)\hat{x} - r\omega^2\sin(\omega t)\hat{y}$$

$$\mathbf{a} = -\omega^2(r\cos(\omega t)\hat{x} + r\sin(\omega t)\hat{y})$$

Notice that the term in parenthesis is the original expression of $\mathbf{r}$ in Cartesian coordinates. Consequently,

$$\mathbf{a} = -\omega^2\mathbf{r}.$$

negative shows that the acceleration is pointed towards the center of the circle (opposite the radius), hence it is called "centripetal" (i.e. "center-seeking"). While objects naturally follow a straight path (due to inertia), this centripetal acceleration describes the circular motion path caused by a centripetal force.

Derivation using Vectors

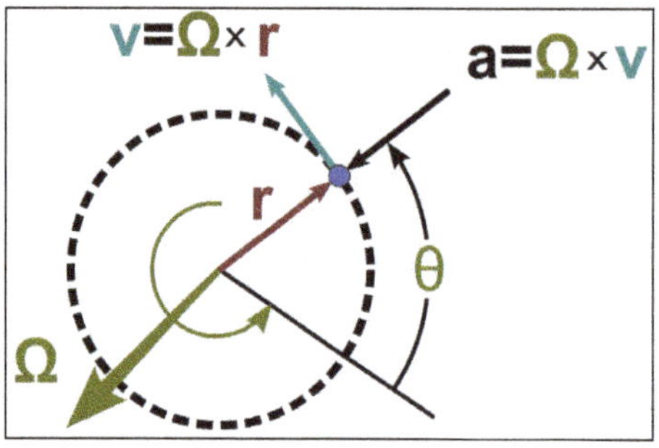

Vector relationships for uniform circular motion; vector Ω representing the rotation is normal to the plane of the orbit with polarity determined by the right-hand rule and magnitude $d\theta/dt$.

The image above shows the vector relationships for uniform circular motion. The rotation itself is represented by the angular velocity vector Ω, which is normal to the plane of the orbit (using the right-hand rule) and has magnitude given by:

$$\Omega| = \frac{\mathrm{d}\theta}{\mathrm{d}t} = \omega,$$

with θ the angular position at time t. In this subsection, $\mathrm{d}\theta/\mathrm{d}t$ is assumed constant, independent of time. The distance traveled $\mathrm{d}\ell$ of the particle in time $\mathrm{d}t$ along the circular path is,

$$\mathrm{d}\ell = \Omega \times \mathbf{r}(t)\mathrm{d}t,$$

which, by properties of the vector cross product, has magnitude $r\mathrm{d}\theta$ and is in the direction tangent to the circular path.

Consequently,

$$\frac{\mathrm{d}\mathbf{r}}{\mathrm{d}t} = \lim_{\Delta t \to 0} \frac{\mathbf{r}(t+\Delta t) - \mathbf{r}(t)}{\Delta t} = \frac{\mathrm{d}\ell}{\mathrm{d}t}.$$

In other words,

$$\mathbf{v} \overset{\text{def}}{=} \frac{\mathrm{d}\mathbf{r}}{\mathrm{d}t} = \frac{\mathrm{d}\ell}{\mathrm{d}t} = \Omega \times \mathbf{r}(t).$$

Differentiating with respect to time,

$$\mathbf{a} \overset{\text{def}}{=} \frac{\mathrm{d}\mathbf{v}}{dt} = \Omega \times \frac{\mathrm{d}\mathbf{r}(t)}{\mathrm{d}t} = \Omega \times [\Omega \times \mathbf{r}(t)].$$

Lagrange's formula states:

$$\mathbf{a} \times (\mathbf{b} \times \mathbf{c}) = \mathbf{b}(\mathbf{a} \cdot \mathbf{c}) - \mathbf{c}(\mathbf{a} \cdot \mathbf{b}).$$

Applying Lagrange's formula with the observation that $\Omega \bullet \mathrm{r}(t) = 0$ at all times,

$$\mathbf{a} = -|\Omega|^2 \mathbf{r}(t).$$

In words, the acceleration is pointing directly opposite to the radial displacement r at all times, and has a magnitude:

$$|\mathbf{a}| = |\mathbf{r}(t)| \left(\frac{d\theta}{dt} \right)^2 = r\omega^2$$

where vertical bars |...| denote the vector magnitude, which in the case of r(t) is simply the radius r of the path.

When the rate of rotation is made constant in the analysis of nonuniform circular motion, that analysis agrees with this one.

A merit of the vector approach is that it is manifestly independent of any coordinate system.

Example: The Banked Turn

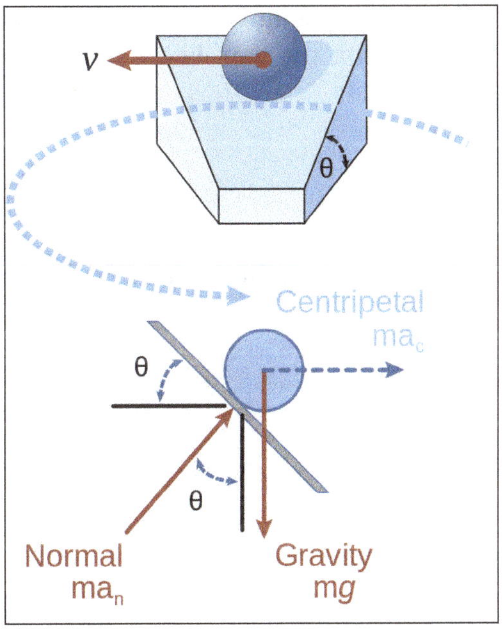

Upper panel: Ball on a banked circular track moving with constant speed v; Lower panel: Forces on the ball.

The upper panel in the image at right shows a ball in circular motion on a banked curve. The curve is banked at an angle θ from the horizontal, and the surface of the road is considered to be slippery. The objective is to find what angle the bank must have so the ball does not slide off the road. Intuition tells us that, on a flat curve with no banking at all, the ball will simply slide off the road; while with a very steep banking, the ball will slide to the center unless it travels the curve rapidly.

Apart from any acceleration that might occur in the direction of the path, the lower panel of the image above indicates the forces on the ball. There are *two* forces; one is the force of gravity vertically downward through the center of mass of the ball mg, where m is the mass of the ball and g is the gravitational acceleration; the second is the upward normal force exerted by the road at a right angle to the road surface ma_n. The centripetal force demanded by the curved motion is also shown above. This centripetal force is not a third force applied to the ball, but rather must be provided by the net force on the ball resulting from vector addition of the normal force and the force of gravity. The resultant or net force on the ball found by vector addition of the normal force exerted by the road and vertical force due to gravity must equal the centripetal force dictated by the need to travel a circular path. The curved motion is maintained so long as this net force provides the centripetal force requisite to the motion.

The horizontal net force on the ball is the horizontal component of the force from the road, which has magnitude $|F_h| = m|a_n|\sin\theta$. The vertical component of the force from the road must counteract the gravitational force: $|F_v| = m|a_n|\cos\theta = m|g|$, which implies $|a_n|=|g| / \cos\theta$. Substituting into the above formula for $|F_h|$ yields a horizontal force to be:

$$|\mathbf{F}_h| = m|\mathbf{g}|\frac{\sin\theta}{\cos\theta} = m|\mathbf{g}|\tan\theta.$$

On the other hand, at velocity $|v|$ on a circular path of radius r, kinematics says that the force needed to turn the ball continuously into the turn is the radially inward centripetal force F_c of magnitude:

$$|\mathbf{F}_c| = m|\mathbf{a}_c| = \frac{m|\mathbf{v}|^2}{r}.$$

Consequently, the ball is in a stable path when the angle of the road is set to satisfy the condition:

$$m|\mathbf{g}|\tan\theta = \frac{m|\mathbf{v}|^2}{r},$$

or,

$$\tan\theta = \frac{|\mathbf{v}|^2}{|\mathbf{g}|r}.$$

As the angle of bank θ approaches 90°, the tangent function approaches infinity, allowing larger values for $|v|^2/r$. In words, this equation states that for greater speeds (bigger $|v|$) the road must be banked more steeply (a larger value for θ), and for sharper turns (smaller r) the road also must be banked more steeply, which accords with intuition. When the angle θ does not satisfy the above condition, the horizontal component of force exerted by the road does not provide the correct centripetal force, and an additional frictional force tangential to the road surface is called upon to provide the difference. If friction cannot do this (that is, the coefficient of friction is exceeded), the ball slides to a different radius where the balance can be realized.

Nonuniform Circular Motion

As a generalization of the uniform circular motion case, suppose the angular rate of rotation is not constant. The acceleration now has a tangential component, as shown the image at right. This case is used to demonstrate a derivation strategy based on a polar coordinate system.

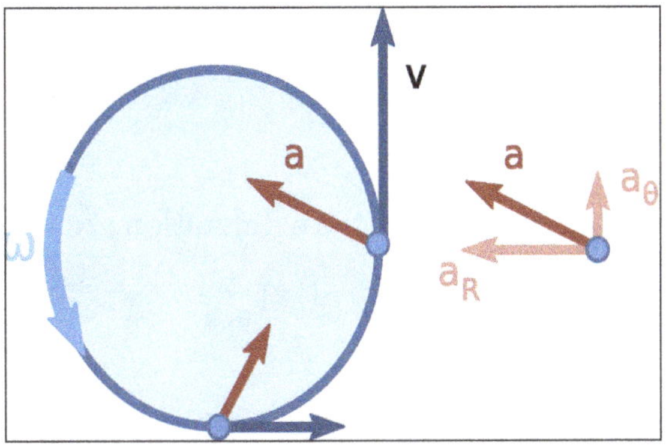

Velocity and acceleration for nonuniform circular motion: the velocity vector is tangential to the orbit, but the acceleration vector is not radially inward because of its tangential component a_θ that increases the rate of rotation: $d\omega / dt = | a_\theta | / R$.

Let r(t) be a vector that describes the position of a point mass as a function of time. Since we are assuming circular motion, let $\mathbf{r}(t) = R\cdot\mathbf{u}_r$, where R is a constant (the radius of the circle) and u_r is the unit vector pointing from the origin to the point mass. The direction of $\mathbf{u}_r$ is described by θ, the angle between the x-axis and the unit vector, measured counterclockwise from the x-axis. The other unit vector for polar coordinates, $\mathbf{u}_\theta$ is perpendicular to $\mathbf{u}_r$ and points in the direction of increasing θ. These polar unit vectors can be expressed in terms of Cartesian unit vectors in the x and y directions, denoted **i** and **j** respectively:

$$\mathbf{ur} = \cos\theta\,\mathbf{i} + \sin\theta\,\mathbf{j}$$

and

$$\mathbf{u}_\theta = -\sin\theta\,\mathbf{i} + \cos\theta\,\mathbf{j}.$$

One can differentiate to find velocity:

$$\mathbf{v} = r\frac{d\mathbf{u}_r}{dt} = r\frac{d}{dt}(\cos\theta\,\mathbf{i} + \sin\theta\,\mathbf{j})$$

$$= r\frac{d\theta}{dt}(-\sin\theta\,\mathbf{i} + \cos\theta\,\mathbf{j})$$

$$= r\frac{d\theta}{dt}\mathbf{u}_\theta$$

$$= \omega r\mathbf{u}_\theta$$

where ω is the angular velocity $d\theta/dt$.

This result for the velocity matches expectations that the velocity should be directed tangentially to the circle, and that the magnitude of the velocity should be $r\omega$. Differentiating again, and noting that:

$$\frac{d\mathbf{u}_\theta}{dt} = -\frac{d\theta}{dt}\mathbf{u}_r = -\omega\mathbf{u}_r,$$

we find that the acceleration, a is:

$$\mathbf{a} = r\left(\frac{\mathrm{d}\omega}{\mathrm{d}t}\mathbf{u}_\theta - \omega^2\mathbf{u}_\mathrm{r}\right).$$

Thus, the radial and tangential components of the acceleration are:

$$\mathbf{a}_\mathrm{r} = -\omega^2 r\,\mathbf{u}_\mathrm{r} = -\frac{|\mathbf{v}|^2}{r}\mathbf{u}_\mathrm{r} \text{ and } \mathbf{a}_\theta = r\frac{\mathrm{d}\omega}{\mathrm{d}t}\mathbf{u}_\theta = \frac{\mathrm{d}|\mathbf{v}|}{\mathrm{d}t}\mathbf{u}_\theta\,,$$

where $|v| = r\,\omega$ is the magnitude of the velocity (the speed).

These equations express mathematically that, in the case of an object that moves along a circular path with a changing speed, the acceleration of the body may be decomposed into a perpendicular component that changes the direction of motion (the centripetal acceleration), and a parallel, or tangential component, that changes the speed.

General Planar Motion

Polar Coordinates

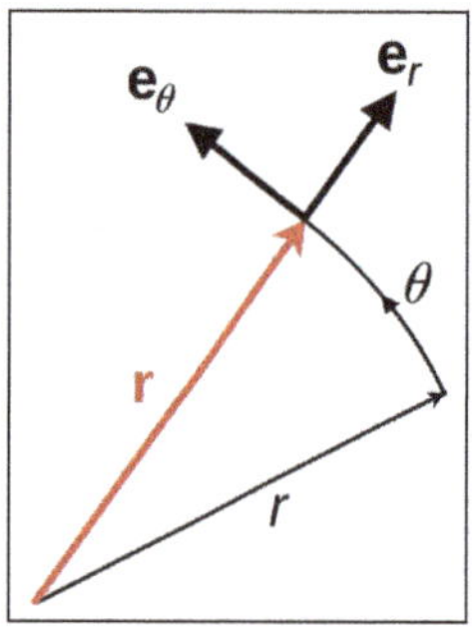

Position vector r, always points radially from the origin.

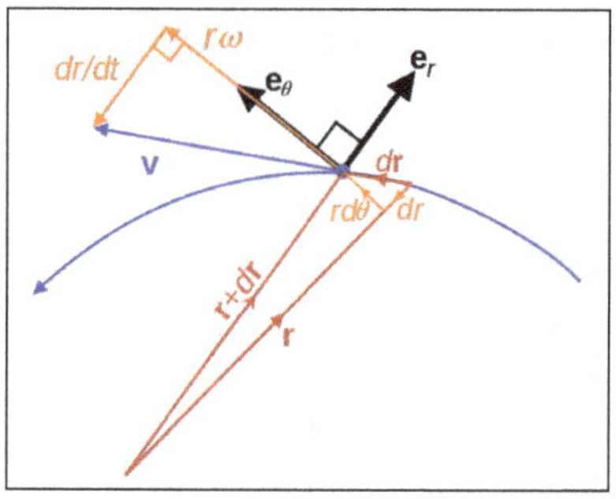

Velocity vector v, always tangent to the path of motion.

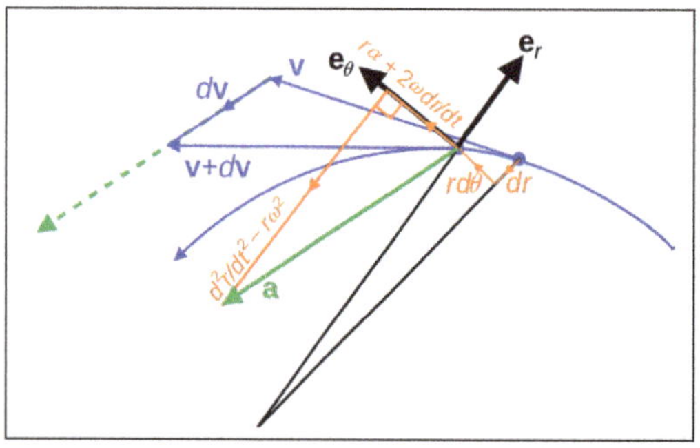

Acceleration vector a, not parallel to the radial motion but offset by the angular and Coriolis accelerations, nor tangent to the path but offset by the centripetal and radial accelerations.

Kinematic vectors in plane polar coordinates. Notice the setup is not restricted to 2d space, but a plane in any higher dimension.

The above results can be derived perhaps more simply in polar coordinates, and at the same time extended to general motion within a plane, as shown next. Polar coordinates in the plane employ a radial unit vector $\mathbf{u}_\rho$ and an angular unit vector $\mathbf{u}_\theta$, as shown above. A particle at position $\mathbf{r}$ is described by:

$$\mathbf{r} = \rho \mathbf{u}_\rho ,$$

where the notation $\boldsymbol{\rho}$ is used to describe the distance of the path from the origin instead of R to emphasize that this distance is not fixed, but varies with time. The unit vector $\mathbf{u}_\rho$ travels with the particle and always points in the same direction as r(t). Unit vector $\mathbf{u}_\theta$ also travels with the particle and stays orthogonal to $\mathbf{u}_\rho$. Thus, $\mathbf{u}_\rho$ and $\mathbf{u}_\theta$ form a local Cartesian coordinate system attached to the particle, and tied to the path traveled by the particle. By moving the unit vectors so their tails coincide, as seen in the circle at the left of the image above, it is seen that $\mathbf{u}_\rho$ and $\mathbf{u}_\theta$ form a right-angled pair with tips on the unit circle that trace back and forth on the perimeter of this circle with the same angle $\theta(t)$ as $\mathbf{r}(t)$.

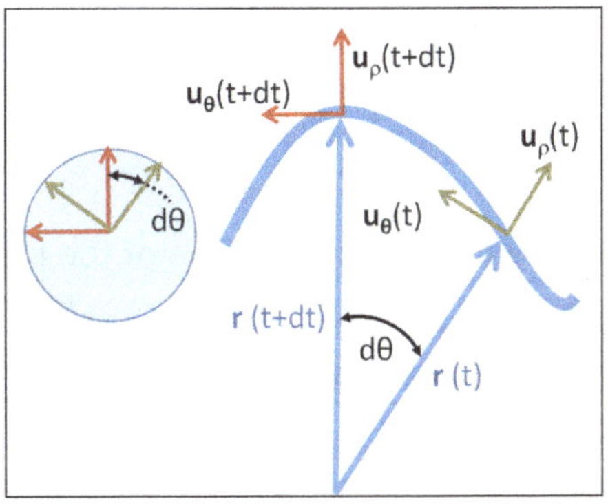

Polar unit vectors, in the figure above, at two times t and $t + dt$ for a particle with trajectory $\mathbf{r}$ (t); on the left the unit vectors $\mathbf{u}_\rho$ and $\mathbf{u}_\theta$ at the two times are moved so their tails all meet, and are shown to trace an arc of a unit radius circle. Their rotation in time dt is $d\theta$, just the same angle as the rotation of the trajectory $\mathbf{r}$ (t).

When the particle moves, its velocity is,

$$\mathbf{v} = \frac{d\rho}{dt}\mathbf{u}_\rho + \rho \frac{d\mathbf{u}_\rho}{dt} .$$

To evaluate the velocity, the derivative of the unit vector $\mathbf{u}_\rho$ is needed. Because $\mathbf{u}_\rho$ is a unit vector, its magnitude is fixed, and it can change only in direction, that is, its change d$\mathbf{u}_\rho$ has a component only perpendicular to $\mathbf{u}_\rho$. When the trajectory $\mathbf{r}(t)$ rotates an amount dθ, $\mathbf{u}_\rho$, which points in the same direction as $\mathbf{r}(t)$, also rotates by dθ. See image above. Therefore, the change in $\mathbf{u}_\rho$ is,

$$d\mathbf{u}_\rho = \mathbf{u}_\theta d\theta ,$$

or

$$\frac{d\mathbf{u}_\rho}{dt} = \mathbf{u}_\theta \frac{d\theta}{dt} .$$

In a similar fashion, the rate of change of u_θ is found. As with $\mathbf{u}_\rho$, $\mathbf{u}_\theta$ is a unit vector and can only rotate without changing size. To remain orthogonal to $\mathbf{u}_\rho$ while the trajectory $\mathbf{r}(t)$ rotates an amount

$d\theta$, $\mathbf{u}_\theta$, which is orthogonal to $\mathbf{r}(t)$, also rotates by $d\theta$. Therefore, the change $d\mathbf{u}_\theta$ is orthogonal to $\mathbf{u}_\theta$ and proportional to $d\theta$:

$$\frac{d\mathbf{u}_\theta}{dt} = -\frac{d\theta}{dt}\mathbf{u}_\rho .$$

The image above shows the sign to be negative: to maintain orthogonality, if $d\mathbf{u}_\rho$ is positive with $d\theta$, then du_θ must decrease.

Substituting the derivative of $\mathbf{u}_\rho$ into the expression for velocity:

$$\mathbf{v} = \frac{d\rho}{dt}\mathbf{u}_\rho + \rho\mathbf{u}_\theta\frac{d\theta}{dt} = v_\rho\mathbf{u}_\rho + v_\theta\mathbf{u}_\theta = \mathbf{v}_\rho + \mathbf{v}_\theta .$$

To obtain the acceleration, another time differentiation is done:

$$\mathbf{a} = \frac{d^2\rho}{dt^2}\mathbf{u}_\rho + \frac{d\rho}{dt}\frac{d\mathbf{u}_\rho}{dt} + \frac{d\rho}{dt}\mathbf{u}_\theta\frac{d\theta}{dt} + \rho\frac{d\mathbf{u}_\theta}{dt}\frac{d\theta}{dt} + \rho\mathbf{u}_\theta\frac{d^2\theta}{dt^2} .$$

Substituting the derivatives of u_ρ and u_θ, the acceleration of the particle is:

$$\mathbf{a} = \frac{d^2\rho}{dt^2}\mathbf{u}_\rho + 2\frac{d\rho}{dt}\mathbf{u}_\theta\frac{d\theta}{dt} - \rho\mathbf{u}_\rho\left(\frac{d\theta}{dt}\right)^2 + \rho\mathbf{u}_\theta\frac{d^2\theta}{dt^2} ,$$

$$= \mathbf{u}_\rho\left[\frac{d^2\rho}{dt^2} - \rho\left(\frac{d\theta}{dt}\right)^2\right] + \mathbf{u}_\theta\left[2\frac{d\rho}{dt}\frac{d\theta}{dt} + \rho\frac{d^2\theta}{dt^2}\right]$$

$$= \mathbf{u}_\rho\left[\frac{dv_\rho}{dt} - \frac{v_\theta^2}{\rho}\right] + \mathbf{u}_\theta\left[\frac{2}{\rho}v_\rho v_\theta + \rho\frac{d}{dt}\frac{v_\theta}{\rho}\right].$$

As a particular example, if the particle moves in a circle of constant radius R, then $d\rho/dt = 0$, $\mathbf{v} = \mathbf{v}_\theta$, and:

$$\mathbf{a} = \mathbf{u}_\rho\left[-\rho\left(\frac{d\theta}{dt}\right)^2\right] + \mathbf{u}_\theta\left[\rho\frac{d^2\theta}{dt^2}\right]$$

$$= \mathbf{u}_\rho\left[-\frac{v^2}{r}\right] + \mathbf{u}_\theta\left[\frac{dv}{dt}\right]$$

where $v = v_\theta$.

These results agree with those above for nonuniform circular motion.If this acceleration is multiplied by the particle mass, the leading term is the centripetal force and the negative of the second term related to angular acceleration is sometimes called the Euler force.

For trajectories other than circular motion, for example, the more general trajectory envisioned in the image above, the instantaneous center of rotation and radius of curvature of the trajectory

are related only indirectly to the coordinate system defined by $\mathbf{u}_\rho$ and $\mathbf{u}_\theta$ and to the length $|\mathbf{r}(t)| = \rho$. Consequently, in the general case, it is not straightforward to disentangle the centripetal and Euler terms from the above general acceleration equation. To deal directly with this issue, local coordinates are preferable.

Local Coordinates

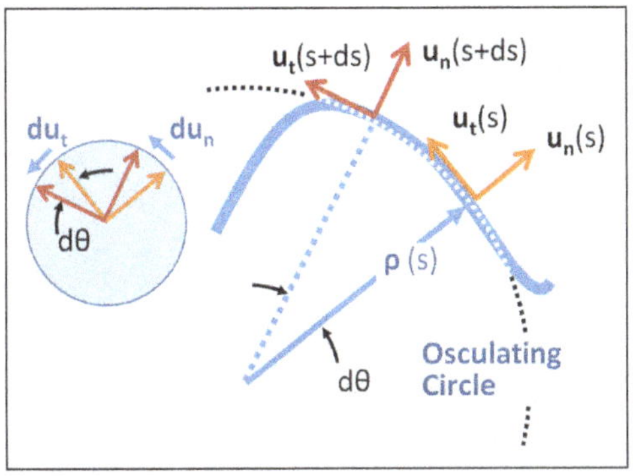

Local coordinate system for planar motion on a curve.

Two different positions are shown in the figure above for distances s and $s + ds$ along the curve. At each position s, unit vector $\mathbf{u}_n$ points along the outward normal to the curve and unit vector $\mathbf{u}_t$ is tangential to the path. The radius of curvature of the path is ρ as found from the rate of rotation of the tangent to the curve with respect to arc length, and is the radius of the osculating circle at position s. The unit circle on the left shows the rotation of the unit vectors with s.

Local coordinates mean a set of coordinates that travel with the particle, and have orientation determined by the path of the particle. Unit vectors are formed as shown in the figure above, both tangential and normal to the path. This coordinate system sometimes is referred to as intrinsic or path coordinates or nt-coordinates, for normal-tangential, referring to these unit vectors. These coordinates are a very special example of a more general concept of local coordinates from the theory of differential forms.

Distance along the path of the particle is the arc length s, considered to be a known function of time.

$$s = s(t).$$

A center of curvature is defined at each position s located a distance ρ (the radius of curvature) from the curve on a line along the normal $u_n(s)$. The required distance $\rho(s)$ at arc length s is defined in terms of the rate of rotation of the tangent to the curve, which in turn is determined by the path itself. If the orientation of the tangent relative to some starting position is $\theta(s)$, then $\rho(s)$ is defined by the derivative $\mathrm{d}\theta/\mathrm{d}s$:

$$\frac{1}{\rho(s)} = \kappa(s) = \frac{\mathrm{d}\theta}{\mathrm{d}s}.$$

The radius of curvature usually is taken as positive (that is, as an absolute value), while the *curvature* κ is a signed quantity.

A geometric approach to finding the center of curvature and the radius of curvature uses a limiting process leading to the osculating circle.

Using these coordinates, the motion along the path is viewed as a succession of circular paths of ever-changing center, and at each position s constitutes non-uniform circular motion at that position with radius ρ. The local value of the angular rate of rotation then is given by:

$$\omega(s) = \frac{d\theta}{dt} = \frac{d\theta}{ds}\frac{ds}{dt} = \frac{1}{\rho(s)}\frac{ds}{dt} = \frac{v(s)}{\rho(s)},$$

with the local speed v given by:

$$v(s) = \frac{ds}{dt}.$$

As for the other examples above, because unit vectors cannot change magnitude, their rate of change is always perpendicular to their direction:

$$\frac{d\mathbf{u}_n(s)}{ds} = \mathbf{u}_t(s)\frac{d\theta}{ds} = \mathbf{u}_t(s)\frac{1}{\rho}\ ; \quad \frac{d\mathbf{u}_t(s)}{ds} = -\mathbf{u}_n(s)\frac{d\theta}{ds} = -\mathbf{u}_n(s)\frac{1}{\rho}.$$

Consequently, the velocity and acceleration are:

$$\mathbf{v}(t) = v\mathbf{u}_t(s);$$

and using the chain-rule of differentiation:

$$\mathbf{a}(t) = \frac{dv}{dt}\mathbf{u}_t(s) - \frac{v^2}{\rho}\mathbf{u}_n(s);\ \text{with the tangential acceleration } \frac{dv}{dt} = \frac{dv}{ds}\frac{ds}{dt} = \frac{dv}{ds}v.$$

In this local coordinate system, the acceleration resembles the expression for nonuniform circular motion with the local radius $\rho(s)$, and the centripetal acceleration is identified as the second term.

Extending this approach to three dimensional space curves leads to the Frenet–Serret formulas.

Alternative Approach

Looking at the image above, one might wonder whether adequate account has been taken of the difference in curvature between $\rho(s)$ and $\rho(s + ds)$ in computing the arc length as $ds = \rho(s)d\theta$.

To introduce the unit vectors of the local coordinate system, one approach is to begin in Cartesian coordinates and describe the local coordinates in terms of these Cartesian coordinates. In terms of arc length s, let the path be described as:

$$\mathbf{r}(s) = [x(s), y(s)].$$

Then an incremental displacement along the path ds is described by:

$$d\mathbf{r}(s) = [dx(s), dy(s)] = [x'(s), y'(s)]ds,$$

where primes are introduced to denote derivatives with respect to *s*. The magnitude of this displacement is d*s*, showing that:

$$\left[x'(s)^2 + y'(s)^2\right] = 1.$$

This displacement is necessarily a tangent to the curve at *s*, showing that the unit vector tangent to the curve is:

$$\mathbf{u}_t(s) = \left[x'(s), y'(s)\right],$$

while the outward unit vector normal to the curve is,

$$\mathbf{u}_n(s) = \left[y'(s), -x'(s)\right],$$

Orthogonality can be verified by showing that the vector dot product is zero. The unit magnitude of these vectors is a consequence of the equation above. Using the tangent vector, the angle θ of the tangent to the curve is given by:

$$\sin\theta = \frac{y'(s)}{\sqrt{x'(s)^2 + y'(s)^2}} = y'(s)\text{; and } \cos\theta = \frac{x'(s)}{\sqrt{x'(s)^2 + y'(s)^2}} = x'(s).$$

The radius of curvature is introduced completely formally (without need for geometric interpretation) as:

$$\frac{1}{\rho} = \frac{d\theta}{ds}.$$

The derivative of θ can be found from that for $\sin\theta$:

$$\frac{d\sin\theta}{ds} = \cos\theta\frac{d\theta}{ds} = \frac{1}{\rho}\cos\theta = \frac{1}{\rho}x'(s).$$

Now:

$$\frac{d\sin\theta}{ds} = \frac{d}{ds}\frac{y'(s)}{\sqrt{x'(s)^2 + y'(s)^2}}$$

$$= \frac{y''(s)x'(s)^2 - y'(s)x'(s)x''(s)}{\left(x'(s)^2 + y'(s)^2\right)^{3/2}},$$

in which the denominator is unity. With this formula for the derivative of the sine, the radius of curvature becomes:

$$\frac{d\theta}{ds} = \frac{1}{\rho} = y''(s)x'(s) - y'(s)x''(s) = \frac{y''(s)}{x'(s)} = -\frac{x''(s)}{y'(s)},$$

where the equivalence of the forms stems from differentiation of $\left[x'(s)^2 + y'(s)^2\right] = 1$:

$$x'(s)x''(s) + y'(s)y''(s) = 0.$$

With these results, the acceleration can be found:

$$\mathbf{a}(s) = \frac{\mathrm{d}}{\mathrm{d}t}\mathbf{v}(s) = \frac{\mathrm{d}}{\mathrm{d}t}\left[\frac{\mathrm{d}s}{\mathrm{d}t}(x'(s), y'(s))\right]$$

$$= \left(\frac{\mathrm{d}^2 s}{\mathrm{d}t^2}\right)\mathbf{u}_\mathrm{t}(s) + \left(\frac{\mathrm{d}s}{\mathrm{d}t}\right)^2 (x''(s), y''(s))$$

$$= \left(\frac{\mathrm{d}^2 s}{\mathrm{d}t^2}\right)\mathbf{u}_\mathrm{t}(s) - \left(\frac{\mathrm{d}s}{\mathrm{d}t}\right)^2 \frac{1}{\rho}\mathbf{u}_\mathrm{n}(s),$$

as can be verified by taking the dot product with the unit vectors $\mathbf{u}_\mathrm{t}(s)$ and $\mathbf{u}_\mathrm{n}(s)$. This result for acceleration is the same as that for circular motion based on the radius ρ. Using this coordinate system in the inertial frame, it is easy to identify the force normal to the trajectory as the centripetal force and that parallel to the trajectory as the tangential force. From a qualitative standpoint, the path can be approximated by an arc of a circle for a limited time, and for the limited time a particular radius of curvature applies, the centrifugal and Euler forces can be analyzed on the basis of circular motion with that radius.

This result for acceleration agrees with that found earlier. However, in this approach, the question of the change in radius of curvature with *s* is handled completely formally, consistent with a geometric interpretation, but not relying upon it, thereby avoiding any questions the image above might suggest about neglecting the variation in ρ.

Example: Circular Motion

To illustrate the above formulas, let *x*, *y* be given as:

$$x = \alpha \cos\frac{s}{\alpha}\,;\ y = \alpha \sin\frac{s}{\alpha}.$$

Then:

$$x^2 + y^2 = \alpha^2,$$

which can be recognized as a circular path around the origin with radius α. The position $s = 0$ corresponds to $[\alpha, 0]$, or 3 o'clock. To use the above formalism, the derivatives are needed:

$$y'(s) = \cos\frac{s}{\alpha}\,;\ x'(s) = -\sin\frac{s}{\alpha},$$

$$y''(s) = -\frac{1}{\alpha}\sin\frac{s}{\alpha}\,;\ x''(s) = -\frac{1}{\alpha}\cos\frac{s}{\alpha}.$$

With these results, one can verify that:

$$x'(s)^2 + y'(s)^2 = 1\,;\ \frac{1}{\rho} = y''(s)x'(s) - y'(s)x''(s) = \frac{1}{\alpha}.$$

The unit vectors can also be found:

$$\mathbf{u}_t(s) = \left[-\sin\frac{s}{\alpha}, \cos\frac{s}{\alpha}\right]; \mathbf{u}_n(s) = \left[\cos\frac{s}{\alpha}, \sin\frac{s}{\alpha}\right],$$

which serve to show that $s = 0$ is located at position $[\rho, 0]$ and $s = \rho\pi/2$ at $[0, \rho]$, which agrees with the original expressions for x and y. In other words, s is measured counterclockwise around the circle from 3 o'clock. Also, the derivatives of these vectors can be found:

$$\frac{d}{ds}\mathbf{u}_t(s) = -\frac{1}{\alpha}\left[\cos\frac{s}{\alpha}, \sin\frac{s}{\alpha}\right] = -\frac{1}{\alpha}\mathbf{u}_n(s);$$

$$\frac{d}{ds}\mathbf{u}_n(s) = \frac{1}{\alpha}\left[-\sin\frac{s}{\alpha}, \cos\frac{s}{\alpha}\right] = \frac{1}{\alpha}\mathbf{u}_t(s).$$

To obtain velocity and acceleration, a time-dependence for s is necessary. For counterclockwise motion at variable speed $v(t)$:

$$s(t) = \int_0^t dt'\ v(t'),$$

where $v(t)$ is the speed and t is time, and $s(t = 0) = 0$. Then:

$$\mathbf{v} = v(t)\mathbf{u}_t(s),$$

$$\mathbf{a} = \frac{dv}{dt}\mathbf{u}_t(s) + v\frac{d}{dt}\mathbf{u}_t(s) = \frac{dv}{dt}\mathbf{u}_t(s) - v\frac{1}{\alpha}\mathbf{u}_n(s)\frac{ds}{dt}$$

$$\mathbf{a} = \frac{dv}{dt}\mathbf{u}_t(s) - \frac{v^2}{\alpha}\mathbf{u}_n(s),$$

where it already is established that $\alpha = \rho$. This acceleration is the standard result for non-uniform circular motion.

Centrifugal Force

In Newtonian mechanics, the centrifugal force is an inertial force (also called a "fictitious" or "pseudo" force) that appears to act on all objects when viewed in a rotating frame of reference. It is directed away from an axis passing through the coordinate system's origin and parallel to the axis of rotation. If the axis of rotation passes through the coordinate system's origin, the centrifugal force is directed radially outwards from that axis. The concept of centrifugal force can be applied in rotating devices, such as centrifuges, centrifugal pumps, centrifugal governors, and centrifugal clutches, and in centrifugal railways, planetary orbits and banked curves, when they are analyzed in a rotating coordinate system. The term has sometimes also been used for the reactive centrifugal force that may be viewed as a reaction to a centripetal force in some circumstances.

Centrifugal force is an outward force apparent in a rotating reference frame. It does not exist when a system is described relative to an inertial frame of reference.

All measurements of position and velocity must be made relative to some frame of reference. For example, an analysis of the motion of an object in an airliner in flight could be made relative to the airliner, to the surface of the Earth, or even to the Sun. A reference frame that is at rest (or one that moves with no rotation and at constant velocity) relative to the "fixed stars" is generally taken to be an inertial frame. Any system can be analyzed in an inertial frame (and so with no centrifugal force). However, it is often more convenient to describe a rotating system by using a rotating frame—the calculations are simpler, and descriptions more intuitive. When this choice is made, fictitious forces, including the centrifugal force, arise.

In a reference frame rotating about an axis through its origin, all objects, regardless of their state of motion, appear to be under the influence of a radially (from the axis of rotation) outward force that is proportional to their mass, to the distance from the axis of rotation of the frame, and to the square of the angular velocity of the frame. This is the centrifugal force. As humans usually experience centrifugal force from within the rotating reference frame, e.g. on a merry-go-round or vehicle, this is much more well-known than centripetal force.

Motion relative to a rotating frame results in another fictitious force: the Coriolis force. If the rate of rotation of the frame changes, a third fictitious force (the Euler force) is required. These fictitious forces are necessary for the formulation of correct equations of motion in a rotating reference frame and allow Newton's laws to be used in their normal form in such a frame (with one exception: the fictitious forces do not obey Newton's third law: they have no equal and opposite counterparts).

Examples

Vehicle Driving Round a Curve

A common experience that gives rise to the idea of a centrifugal force is encountered by passengers riding in a vehicle, such as a car, that is changing direction. If a car is traveling at a constant speed along a straight road, then a passenger inside is not accelerating and, according to Newton's second law of motion, the net force acting on him is therefore zero (all forces acting on him cancel each other out). If the car enters a curve that bends to the left, the passenger experiences an apparent force that seems to be pulling him towards the right. This is the fictitious centrifugal force. It is needed within the passenger's local frame of reference to explain his sudden tendency to start accelerating to the right relative to the car—a tendency which he must resist by applying a rightward force to the car (for instance, a frictional force against the seat) in order to remain in a fixed position inside. Since he pushes the seat toward the right, Newton's third law says that the seat pushes him toward the left. The centrifugal force must be included in the passenger's reference frame (in which the passenger remains at rest): it counteracts the leftward force applied to the passenger by the seat, and explains why this otherwise unbalanced force does not cause him to accelerate. However, it would be apparent to a stationary observer watching from an overpass above that the frictional force exerted on the passenger by the seat is not being balanced; it constitutes a net force to the left, causing the passenger to accelerate toward the inside of the curve, as he must in order to keep moving with the car rather than proceeding in a straight line as he otherwise would. Thus the "centrifugal force" he feels is the result of a "centrifugal tendency" caused by inertia. Similar effects are encountered in aeroplanes and roller coasters where the magnitude of the apparent force is often reported in "G's".

Stone on a String

If a stone is whirled round on a string, in a horizontal plane, the only real force acting on the stone in the horizontal plane is applied by the string (gravity acts vertically). There is a net force on the stone in the horizontal plane which acts toward the center.

In an inertial frame of reference, were it not for this net force acting on the stone, the stone would travel in a straight line, according to Newton's first law of motion. In order to keep the stone moving in a circular path, a centripetal force, in this case provided by the string, must be continuously applied to the stone. As soon as it is removed (for example if the string breaks) the stone moves in a straight line. In this inertial frame, the concept of centrifugal force is not required as all motion can be properly described using only real forces and Newton's laws of motion.

In a frame of reference rotating with the stone around the same axis as the stone, the stone is stationary. However, the force applied by the string is still acting on the stone. If one were to apply Newton's laws in their usual (inertial frame) form, one would conclude that the stone should accelerate in the direction of the net applied force—towards the axis of rotation—which it does not do. The centrifugal force and other fictitious forces must be included along with the real forces in order to apply Newton's laws of motion in the rotating frame.

Earth

The Earth constitutes a rotating reference frame because it rotates once every 23 hours and 56 minutes around its axis. Because the rotation is slow, the fictitious forces it produces are often small, and in everyday situations can generally be neglected. Even in calculations requiring high precision, the centrifugal force is generally not explicitly included, but rather lumped in with the gravitational force: the strength and direction of the local "gravity" at any point on the Earth's surface is actually a combination of gravitational and centrifugal forces. However, the fictitious forces can be of arbitrary size. For example, in an Earth-bound reference system, the fictitious force (the net of Coriolis and centrifugal forces) is enormous and is responsible for the sun orbiting around the Earth (in the Earth-bound reference system). This is due to the large mass and velocity of the sun (relative to the Earth).

Weight of an Object at the Poles and on the Equator

If an object is weighed with a simple spring balance at one of the Earth's poles, there are two forces acting on the object: the Earth's gravity, which acts in a downward direction, and the equal and opposite restoring force in the spring, acting upward. Since the object is stationary and not accelerating, there is no net force acting on the object and the force from the spring is equal in magnitude to the force of gravity on the object. In this case, the balance shows the value of the force of gravity on the object.

When the same object is weighed on the equator, the same two real forces act upon the object. However, the object is moving in a circular path as the Earth rotates and therefore experiencing a centripetal acceleration. When considered in an inertial frame (that is to say, one that is not rotating with the Earth), the non-zero acceleration means that force of gravity will not balance with the force from the spring. In order to have a net centripetal force, the magnitude of the restoring force of the spring must be less than the magnitude of force of gravity. Less restoring force in the spring is reflected on the scale as less weight — about 0.3% less at the equator than at the poles. In the Earth reference frame (in which the object being weighed is at rest), the object does not appear to be accelerating, however the two real forces, gravity and the force from the spring, are the same magnitude and do not balance. The centrifugal force must be included to make the sum of the forces be zero to match the apparent lack of acceleration.

In fact, the observed weight difference is more — about 0.53%. Earth's gravity is a bit stronger at the poles than at the equator, because the Earth is not a perfect sphere, so an object at the poles is slightly closer to the center of the Earth than one at the equator; this effect combines with the centrifugal force to produce the observed weight difference.

Equatorial Railway

This thought experiment is more complicated than the previous examples in that it requires the use of the Coriolis force as well as the centrifugal force.

If there were a railway line running round the Earth's equator, a train moving westward along it fast enough would remain stationary in a frame moving (but not rotating) with the Earth; it would stand still as the Earth spun beneath it. In this inertial frame the situation is easy to analyze. The only forces acting on the train (assuming no wind resistance or other horizontal forces) are its gravity (downward) and the equal and opposite (upward) force from the track. There is no net force on the train and it therefore remains stationary.

In a frame rotating with the Earth the train moves in a circular orbit as it travels round the Earth. In this frame, the upward reaction force from the track and the force of gravity on the train remain the same, as they are real forces. However, in the Earth's (rotating) frame, the train is traveling in a circular path and therefore requires a centripetal (downward) force to keep it on this path. Because this uses a rotating frame, the (fictitious) centrifugal force must be applied to the train. This is equal in value to the required centripetal force but acts in an upward direction — the opposite direction to that required. It would seem that there is a net upward force on the train and it should therefore accelerate upward.

The resolution to this paradox lies in the fact that the train is in motion with respect to the rotating frame and is subject to (in addition to the centrifugal force) the Coriolis force, which, in this example, acts in the downward direction and is twice as strong as the centrifugal force.

Derivation

For the following formalism, the rotating frame of reference is regarded as a special case of a non-inertial reference frame that is rotating relative to an inertial reference frame denoted the stationary frame.

Time Derivatives in a Rotating Frame

In a rotating frame of reference, the time derivatives of any vector function $\boldsymbol{P}$ of time—such as the velocity and acceleration vectors of an object—will differ from its time derivatives in the stationary frame. If P_1 P_2, P_3 are the components of $\boldsymbol{P}$ with respect to unit vectors $\boldsymbol{i}, \boldsymbol{j}, \boldsymbol{k}$ directed along the axes of the rotating frame (i.e. $\boldsymbol{P} = P_1\boldsymbol{i} + P_2\boldsymbol{j} + P_3\boldsymbol{k}$), then the first time derivative $[d\boldsymbol{P}/dt]$ of $\boldsymbol{P}$ with respect to the rotating frame is, by definition, $dP_1/dt\,\boldsymbol{i} + dP_2/dt\,\boldsymbol{j} + dP_3/dt\,\boldsymbol{k}$. If the absolute angular velocity of the rotating frame is $\boldsymbol{\omega}$ then the derivative $d\boldsymbol{P}/dt$ of $\boldsymbol{P}$ with respect to the stationary frame is related to $[d\boldsymbol{P}/dt]$ by the equation:

$$\frac{d\mathbf{P}}{dt} = \left[\frac{d\mathbf{P}}{dt}\right] + \boldsymbol{\omega} \times \mathbf{P},$$

where $\times$ denotes the vector cross product. In other words, the rate of change of $\boldsymbol{P}$ in the stationary frame is the sum of its apparent rate of change in the rotating frame and a rate of rotation $\boldsymbol{\omega} \times \mathbf{P}$ attributable to the motion of the rotating frame. The vector $\boldsymbol{\omega}$ has magnitude ω equal to the rate of rotation and is directed along the axis of rotation according to the right-hand rule.

Acceleration

Newton's law of motion for a particle of mass m written in vector form is:

$$\mathbf{F} = m\mathbf{a},$$

where $\boldsymbol{F}$ is the vector sum of the physical forces applied to the particle and $\boldsymbol{a}$ is the absolute acceleration (that is, acceleration in an inertial frame) of the particle, given by:

$$\mathbf{a} = \frac{d^2\mathbf{r}}{dt^2},$$

where $\boldsymbol{r}$ is the position vector of the particle.

By applying the transformation above from the stationary to the rotating frame three times (twice to $\frac{d\mathbf{r}}{dt}$ and once to $\frac{d}{dt}\left[\frac{d\mathbf{r}}{dt}\right]$), the absolute acceleration of the particle can be written as:

$$\begin{aligned}
\mathbf{a} &= \frac{d^2\mathbf{r}}{dt^2} = \frac{d}{dt}\frac{d\mathbf{r}}{dt} = \frac{d}{dt}\left(\left[\frac{d\mathbf{r}}{dt}\right] + \boldsymbol{\omega}\times\mathbf{r}\right) \\
&= \left[\frac{d^2\mathbf{r}}{dt^2}\right] + \boldsymbol{\omega}\times\left[\frac{d\mathbf{r}}{dt}\right] + \frac{d\boldsymbol{\omega}}{dt}\times\mathbf{r} + \boldsymbol{\omega}\times\frac{d\mathbf{r}}{dt} \\
&= \left[\frac{d^2\mathbf{r}}{dt^2}\right] + \boldsymbol{\omega}\times\left[\frac{d\mathbf{r}}{dt}\right] + \frac{d\boldsymbol{\omega}}{dt}\times\mathbf{r} + \boldsymbol{\omega}\times\left(\left[\frac{d\mathbf{r}}{dt}\right] + \boldsymbol{\omega}\times\mathbf{r}\right) \\
&= \left[\frac{d^2\mathbf{r}}{dt^2}\right] + \frac{d\boldsymbol{\omega}}{dt}\times\mathbf{r} + 2\boldsymbol{\omega}\times\left[\frac{d\mathbf{r}}{dt}\right] + \boldsymbol{\omega}\times(\boldsymbol{\omega}\times\mathbf{r}).
\end{aligned}$$

Force

The apparent acceleration in the rotating frame is $\left[\frac{d^2\mathbf{r}}{dt^2}\right]$. An observer unaware of the rotation would expect this to be zero in the absence of outside forces. However, Newton's laws of motion apply only in the inertial frame and describe dynamics in terms of the absolute acceleration $\frac{d^2\mathbf{r}}{dt^2}$.

Therefore, the observer perceives the extra terms as contributions due to fictitious forces. These terms in the apparent acceleration are independent of mass; so it appears that each of these fictitious forces, like gravity, pulls on an object in proportion to its mass. When these forces are added, the equation of motion has the form:

$$\mathbf{F} - m\frac{\mathrm{d}\boldsymbol{\omega}}{\mathrm{d}t}\times\mathbf{r} - 2m\boldsymbol{\omega}\times\left[\frac{\mathrm{d}\mathbf{r}}{\mathrm{d}t}\right] - m\boldsymbol{\omega}\times(\boldsymbol{\omega}\times\mathbf{r}) = m\left[\frac{\mathrm{d}^2\mathbf{r}}{\mathrm{d}t^2}\right].$$

From the perspective of the rotating frame, the additional force terms are experienced just like the real external forces and contribute to the apparent acceleration. The additional terms on the force side of the equation can be recognized as, reading from left to right, the Euler force $-m\mathrm{d}\boldsymbol{\omega}/\mathrm{d}t\times\mathbf{r}$, the Coriolis force $-2m\boldsymbol{\omega}\times[\mathrm{d}\mathbf{r}/\mathrm{d}t]$, and the centrifugal force $-m\boldsymbol{\omega}\times(\boldsymbol{\omega}\times\mathbf{r})$, respectively. Unlike the other two fictitious forces, the centrifugal force always points radially outward from the axis of rotation of the rotating frame, with magnitude $m\omega^2 r$, and unlike the Coriolis force in particular, it is independent of the motion of the particle in the rotating frame. As expected, for a non-rotating inertial frame of reference $(\boldsymbol{\omega}=0)$ the centrifugal force and all other fictitious forces disappear. Similarly, as the centrifugal force is proportional to the distance from object to the axis of rotation of the frame, the centrifugal force vanishes for objects that lie upon the axis.

Absolute Rotation

The interface of two immiscible liquids rotating around a vertical axis is an upward-opening circular paraboloid.

Three scenarios were suggested by Newton to answer the question of whether the absolute rotation of a local frame can be detected; that is, if an observer can decide whether an observed object is rotating or if the observer is rotating.

- The shape of the surface of water rotating in a bucket. The shape of the surface becomes concave to balance the centrifugal force against the other forces upon the liquid.
- The tension in a string joining two spheres rotating about their center of mass. The tension in the string will be proportional to the centrifugal force on each sphere as it rotates around the common center of mass.

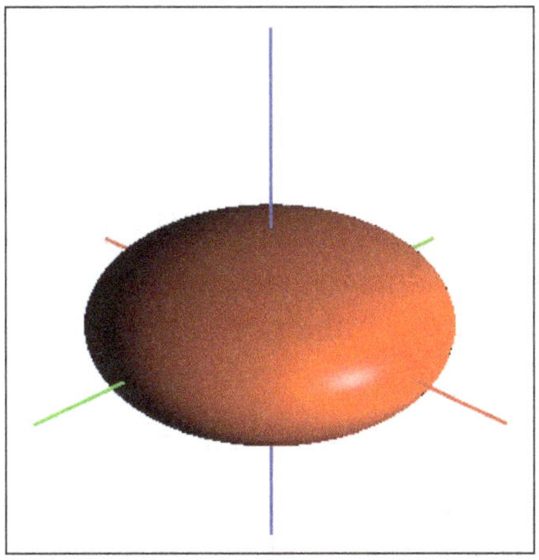

When analysed in a rotating reference frame of the planet, centrifugal force causes rotating planets to assume the shape of an oblate spheroid.

In these scenarios, the effects attributed to centrifugal force are only observed in the local frame (the frame in which the object is stationary) if the object is undergoing absolute rotation relative to an inertial frame. By contrast, in an inertial frame, the observed effects arise as a consequence of the inertia and the known forces without the need to introduce a centrifugal force. Based on this argument, the privileged frame, wherein the laws of physics take on the simplest form, is a stationary frame in which no fictitious forces need to be invoked.

Within this view of physics, any other phenomenon that is usually attributed to centrifugal force can be used to identify absolute rotation. For example, the oblateness of a sphere of freely flowing material is often explained in terms of centrifugal force. The oblate spheroid shape reflects, following Clairaut's theorem, the balance between containment by gravitational attraction and dispersal by centrifugal force. That the Earth is itself an oblate spheroid, bulging at the equator where the radial distance and hence the centrifugal force is larger, is taken as one of the evidences for its absolute rotation.

Applications

The operations of numerous common rotating mechanical systems are most easily conceptualized in terms of centrifugal force. For example:

- A centrifugal governor regulates the speed of an engine by using spinning masses that move radially, adjusting the throttle, as the engine changes speed. In the reference frame of the spinning masses, centrifugal force causes the radial movement.
- A centrifugal clutch is used in small engine-powered devices such as chain saws, go-karts and model helicopters. It allows the engine to start and idle without driving the device but automatically and smoothly engages the drive as the engine speed rises. Inertial drum brake ascenders used in rock climbing and the inertia reels used in many automobile seat belts operate on the same principle.
- Centrifugal forces can be used to generate artificial gravity, as in proposed designs for rotating space stations. The Mars Gravity Biosatellite would have studied the effects of Mars-level gravity on mice with gravity simulated in this way.

- Spin casting and centrifugal casting are production methods that uses centrifugal force to disperse liquid metal or plastic throughout the negative space of a mold.
- Centrifuges are used in science and industry to separate substances. In the reference frame spinning with the centrifuge, the centrifugal force induces a hydrostatic pressure gradient in fluid-filled tubes oriented perpendicular to the axis of rotation, giving rise to large buoyant forces which push low-density particles inward. Elements or particles denser than the fluid move outward under the influence of the centrifugal force. This is effectively Archimedes' principle as generated by centrifugal force as opposed to being generated by gravity.
- Some amusement rides make use of centrifugal forces. For instance, a Gravitron's spin forces riders against a wall and allows riders to be elevated above the machine's floor in defiance of Earth's gravity.

Nevertheless, all of these systems can also be described without requiring the concept of centrifugal force, in terms of motions and forces in a stationary frame, at the cost of taking somewhat more care in the consideration of forces and motions within the system.

Friction

Friction is a force that resists the sliding or rolling of one solid object over another. Frictional forces, such as the traction needed to walk without slipping, may be beneficial, but they also present a great measure of opposition to motion. About 20 percent of the engine power of automobiles is consumed in overcoming frictional forces in the moving parts.

The major cause of friction between metals appears to be the forces of attraction, known as adhesion, between the contact regions of the surfaces, which are always microscopically irregular. Friction arises from shearing these "welded" junctions and from the action of the irregularities of the harder surface plowing across the softer surface.

Two simple experimental facts characterize the friction of sliding solids. First, the amount of friction is nearly independent of the area of contact. If a brick is pulled along a table, the frictional force is the same whether the brick is lying flat or standing on end. Second, friction is proportional to the load or weight that presses the surfaces together. If a pile of three bricks is pulled along a table, the friction is three times greater than if one brick is pulled. Thus, the ratio of friction F to load L is constant. This constant ratio is called the coefficient of friction and is usually symbolized by the Greek letter mu (μ). Mathematically, $\mu = F/L$. Because both friction and load are measured in units of force (such as pounds or newtons), the coefficient of friction is dimensionless. The value of the coefficient of friction for a case of one or more bricks sliding on a clean wooden table is about 0.5, which implies that a force equal to half the weight of the bricks is required just to overcome friction in keeping the bricks moving along at a constant speed. The frictional force itself is directed oppositely to the motion of the object. Because the friction thus far described arises between surfaces in relative motion, it is called kinetic friction.

Static friction, in contrast, acts between surfaces at rest with respect to each other. The value of static friction varies between zero and the smallest force needed to start motion. This smallest force required to start motion, or to overcome static friction, is always greater than the force required to continue the motion, or to overcome kinetic friction.

Rolling friction occurs when a wheel, ball, or cylinder rolls freely over a surface, as in ball and roller bearings. The main source of friction in rolling appears to be dissipation of energy involved in deformation of the objects. If a hard ball is rolling on a level surface, the ball is somewhat flattened and the level surface somewhat indented in the regions in contact. The elastic deformation or compression produced at the leading section of the area in contact is a hindrance to motion that is not fully compensated as the substances spring back to normal shape at the trailing section. The internal losses in the two substances are similar to those that keep a ball from bouncing back to the level from which it is dropped. Coefficients of sliding friction are generally 100 to 1,000 times greater than coefficients of rolling friction for corresponding materials. This advantage was realized historically with the transition from sledge to wheel.

Inertia

Inertia is the resistance of any physical object to any change in its velocity. This includes changes to the object's speed, or direction of motion. An aspect of this property is the tendency of objects to keep moving in a straight line at a constant speed, when no forces act upon them.

Inertia is one of the primary manifestations of mass, which is a quantitative property of physical systems. Isaac Newton defined inertia as his first law in his Philosophiæ Naturalis Principia Mathematica, which states:

> "The *vis insita*, or innate force of matter, is a power of resisting by which every body, as much as in it lies, endeavours to preserve its present state, whether it be of rest or of moving uniformly forward in a straight line."

In common usage, the term "inertia" may refer to an object's "amount of resistance to change in velocity" or for simpler terms, "resistance to a change in motion" (which is quantified by its mass), or sometimes to its momentum, depending on the context. The term "inertia" is more properly understood as shorthand for "the principle of inertia" as described by Newton in his first law of motion: an object not subject to any net external force moves at a constant velocity. Thus, an object will continue moving at its current velocity until some force causes its speed or direction to change.

On the surface of the Earth, inertia is often masked by gravity and the effects of friction and air resistance, both of which tend to decrease the speed of moving objects (commonly to the point of rest). This misled the philosopher Aristotle to believe that objects would move only as long as force was applied to them.

The principle of inertia is one of the fundamental principles in classical physics that are still used today to describe the motion of objects and how they are affected by the applied forces on them.

Rotational Inertia

Another form of inertia is rotational inertia (→ moment of inertia), the property that a rotating rigid body maintains its state of uniform rotational motion. Its angular momentum is unchanged, unless an external torque is applied; this is also called conservation of angular momentum. Rotational inertia depends on the object remaining structurally intact as a rigid body, and also has practical consequences. For example, a gyroscope uses the property that it resists any change in the axis of rotation.

Example: If you had a tablecloth, with food on it, and you pull it quickly, most likely, the objects won't move. That's inertia.

Newton's Laws of Motion

Newton's laws of motion are three physical laws that, together, laid the foundation for classical mechanics. They describe the relationship between a body and the forces acting upon it, and its motion in response to those forces. More precisely, the first law defines the force qualitatively, the second law offers a quantitative measure of the force, and the third asserts that a single isolated force doesn't exist. These three laws have been expressed in several ways, over nearly three centuries, and can be summarised as follows:

- First law: In an inertial frame of reference, an object either remains at rest or continues to move at a constant velocity, unless acted upon by a force.
- Second law: In an inertial frame of reference, the vector sum of the forces **F** on an object is equal to the mass m of that object multiplied by the acceleration a of the object: $\mathbf{F} = m\mathbf{a}$. (It is assumed here that the mass m is constant.)
- Third law: When one body exerts a force on a second body, the second body simultaneously exerts a force equal in magnitude and opposite in direction on the first body.

The three laws of motion were first compiled by Isaac Newton in his Philosophiæ Naturalis Principia Mathematica (Mathematical Principles of Natural Philosophy), first published in 1687. Newton used them to explain and investigate the motion of many physical objects and systems. For example, in the third volume of the text, Newton showed that these laws of motion, combined with his law of universal gravitation, explained Kepler's laws of planetary motion.

Some also describe a fourth law which states that forces add up like vectors, that is, that forces obey the principle of superposition.

Newton's laws are applied to objects which are idealised as single point masses, in the sense that the size and shape of the object's body are neglected to focus on its motion more easily. This can be done when the object is small compared to the distances involved in its analysis, or the deformation and rotation of the body are of no importance. In this way, even a planet can be idealised as a particle for analysis of its orbital motion around a star.

Isaac Newton, the physicist who formulated the laws .

In their original form, Newton's laws of motion are not adequate to characterise the motion of rigid bodies and deformable bodies. Leonhard Euler in 1750 introduced a generalisation of Newton's laws of motion for rigid bodies called Euler's laws of motion, later applied as well for deformable bodies assumed as a continuum. If a body is represented as an assemblage of discrete particles, each governed by Newton's laws of motion, then Euler's laws can be derived from Newton's laws. Euler's laws can, however, be taken as axioms describing the laws of motion for extended bodies, independently of any particle structure.

Newton's laws hold only with respect to a certain set of frames of reference called Newtonian or inertial reference frames. Some authors interpret the first law as defining what an inertial reference frame is; from this point of view, the second law holds only when the observation is made from an inertial reference frame, and therefore the first law cannot be proved as a special case of the second. Other authors do treat the first law as a corollary of the second. The explicit concept of an inertial frame of reference was not developed until long after Newton's death.

In the given interpretation mass, acceleration, momentum, and (most importantly) force are assumed to be externally defined quantities. This is the most common, but not the only interpretation of the way one can consider the laws to be a definition of these quantities.

Newtonian mechanics has been superseded by special relativity, but it is still useful as an approximation when the speeds involved are much slower than the speed of light.

Laws

Newton's laws read:

- Law I: Every body persists in its state of being at rest or of moving uniformly straight forward, except insofar as it is compelled to change its state by force impressed.
- Law II: The alteration of motion is ever proportional to the motive force impress'd; and is made in the direction of the right line in which that force is impress'd.
- Law III: To every action there is always opposed an equal reaction: or the mutual actions of two bodies upon each other are always equal, and directed to contrary parts.

Newton's First Law

The first law states that if the net force (the vector sum of all forces acting on an object) is zero, then the velocity of the object is constant. Velocity is a vector quantity which expresses both the object's speed and the direction of its motion; therefore, the statement that the object's velocity is constant is a statement that both its speed and the direction of its motion are constant.

The first law can be stated mathematically when the mass is a non-zero constant, as,

$$\sum \mathbf{F} = 0 \Leftrightarrow \frac{d\mathbf{v}}{dt} = 0.$$

Consequently,

- An object that is at rest will stay at rest unless a force acts upon it.
- An object that is in motion will not change its velocity unless a force acts upon it.

This is known as uniform motion. An object continues to do whatever it happens to be doing unless a force is exerted upon it. If it is at rest, it continues in a state of rest (demonstrated when a tablecloth is skilfully whipped from under dishes on a tabletop and the dishes remain in their initial state of rest). If an object is moving, it continues to move without turning or changing its speed. This is evident in space probes that continuously move in outer space. Changes in motion must be imposed against the tendency of an object to retain its state of motion. In the absence of net forces, a moving object tends to move along a straight line path indefinitely.

Newton placed the first law of motion to establish frames of reference for which the other laws are applicable. However, Newton implicitly referred to the absolute co-ordinate of cosmos for this frame. Since we cannot precisely measure our velocity relative to a far star, Newton's frame is based on a pure imagination, not based on measurable physics. In current physics, an observer defines himself as in inertial frame by preparing one stone hooked by a spring, and rotating the spring to any direction, and observing the stone static and the length of that spring unchanged. By Einstein's equivalence principle, if there was one such observer A and another observer B moving in a constant velocity related to A, then A and B will both observe the same physics phenomena. if A verified the first law, then B will verify it too. In this way, the definition of inertial can get rid of absolute space or far star, and only refer to the objects locally reachable and measurable.

A particle not subject to forces moves (related to inertial frame) in a straight line at a constant speed. Newton's first law is often referred to as the law of inertia. Thus, a condition necessary for the uniform motion of a particle relative to an inertial reference frame is that the total net force acting on it is zero. In this sense, the first law can be restated as:

> In every material universe, the motion of a particle in a preferential reference frame Φ is determined by the action of forces whose total vanished for all times when and only when the velocity of the particle is constant in Φ. That is, a particle initially at rest or in uniform motion in the preferential frame Φ continues in that state unless compelled by forces to change it.

Newton's first and second laws are valid only in an inertial reference frame. Any reference frame that is in uniform motion with respect to an inertial frame is also an inertial frame, i.e. Galilean invariance or the principle of Newtonian relativity.

Newton's Second Law

The second law states that the rate of change of momentum of a body is directly proportional to the force applied, and this change in momentum takes place in the direction of the applied force.

$$\mathbf{F} = \frac{d\mathbf{p}}{dt} = \frac{d(m\mathbf{v})}{dt}.$$

The second law can also be stated in terms of an object's acceleration. Since Newton's second law is valid only for constant-mass systems, m can be taken outside the differentiation operator by the constant factor rule in differentiation. Thus,

$$\mathbf{F} = m\frac{d\mathbf{v}}{dt} = m\mathbf{a},$$

where F is the net force applied, m is the mass of the body, and a is the body's acceleration. Thus, the net force applied to a body produces a proportional acceleration. In other words, if a body is accelerating, then there is a force on it. An application of this notation is the derivation of G Subscript C.

The above statements hint that the second law is merely a definition of $\mathbf{F}$, not a precious observation of nature. However, current physics restate the second law in measurable steps: (1)defining the term 'one unit of mass' by a specified stone, (2)defining the term 'one unit of force' by a specified spring with specified length, (3)measuring by experiment or proving by theory (with a principle that every direction of space are equivalent), that force can be added as a mathematical vector, (4) finally conclude that $\mathbf{F} = m\mathbf{a}$.. These steps hint the second law is a precious feature of nature.

The second law also implies the conservation of momentum: when the net force on the body is zero, the momentum of the body is constant. Any net force is equal to the rate of change of the momentum.

Any mass that is gained or lost by the system will cause a change in momentum that is not the result of an external force. A different equation is necessary for variable-mass systems.

Newton's second law is an approximation that is increasingly worse at high speeds because of relativistic effects.

According to modern ideas of how Newton was using his terminology, the law is understood, in modern terms, as an equivalent of:

> The change of momentum of a body is proportional to the impulse impressed on the body, and happens along the straight line on which that impulse is impressed.

This may be expressed by the formula F = p', where p' is the time derivative of the momentum p. This equation can be seen clearly in the Wren Library of Trinity College, Cambridge, in a glass case in which Newton's manuscript is open to the relevant page.

Motte's 1729 translation of Newton's Latin continued with Newton's commentary on the second law of motion, reading:

> "If a force generates a motion, a double force will generate double the motion, a triple force triple the motion, whether that force be impressed altogether and at once, or gradually and successively. And this motion (being always directed the same way with the generating force), if the body moved before, is added to or subtracted from the former motion, according as they directly conspire with or are directly contrary to each other; or obliquely joined, when they are oblique, so as to produce a new motion compounded from the determination of both."

The sense or senses in which Newton used his terminology, and how he understood the second law and intended it to be understood, have been extensively discussed by historians of science, along with the relations between Newton's formulation and modern formulations.

Impulse

An impulse J occurs when a force F acts over an interval of time Δt, and it is given by,

$$\mathbf{J} = \int_{\Delta t} \mathbf{F} \mathrm{d}t.$$

Since force is the time derivative of momentum, it follows that

$$\mathbf{J} = \Delta \mathbf{p} = m\Delta \mathbf{v}.$$

This relation between impulse and momentum is closer to Newton's wording of the second law. Impulse is a concept frequently used in the analysis of collisions and impacts.

Variable-mass Systems

Variable-mass systems, like a rocket burning fuel and ejecting spent gases, are not closed and cannot be directly treated by making mass a function of time in the second law; that is, the following formula is wrong:

$$\mathbf{F}_{\text{net}} = \frac{\mathrm{d}}{\mathrm{d}t}\left[m(t)\mathbf{v}(t)\right] = m(t)\frac{\mathrm{d}\mathbf{v}}{\mathrm{d}t} + \mathbf{v}(t)\frac{\mathrm{d}m}{\mathrm{d}t}. \qquad \text{(wrong)}$$

The falsehood of this formula can be seen by noting that it does not respect Galilean invariance: a variable-mass object with F = 0 in one frame will be seen to have F ≠ 0 in another frame. The correct equation of motion for a body whose mass m varies with time by either ejecting or accreting mass is obtained by applying the second law to the entire, constant-mass system consisting of the body and its ejected/accreted mass; the result is,

$$\mathbf{F} + \mathbf{u}\frac{\mathrm{d}m}{\mathrm{d}t} = m\frac{\mathrm{d}\mathbf{v}}{\mathrm{d}t}$$

where u is the velocity of the escaping or incoming mass relative to the body. From this equation one can derive the equation of motion for a varying mass system, for example, the Tsiolkovsky rocket equation. Under some conventions, the quantity $\mathbf{u}\,\mathrm{d}m/\mathrm{d}t$ on the left-hand side, which represents the advection of momentum, is defined as a force (the force exerted on the body by the changing mass, such as rocket exhaust) and is included in the quantity F. Then, by substituting the definition of acceleration, the equation becomes F = *ma*.

Newton's Third Law

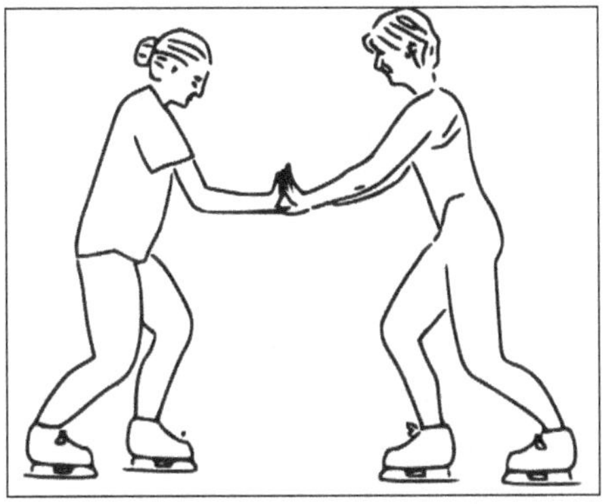

An illustration of Newton's third law in which two skaters push against each other. The first skater on the left exerts a normal force N_{12} on the second skater directed towards the right, and the second skater exerts a normal force N_{21} on the first skater directed towards the left.

The magnitudes of both forces are equal, but they have opposite directions, as dictated by Newton's third law.

The third law states that all forces between two objects exist in equal magnitude and opposite direction: if one object A exerts a force F_A on a second object B, then B simultaneously exerts a force F_B on A, and the two forces are equal in magnitude and opposite in direction: $F_A = -F_B$. The third law means that all forces are interactions between different bodies, or different regions within one body, and thus that there is no such thing as a force that is not accompanied by an equal and opposite force. In some situations, the magnitude and direction of the forces are determined entirely by one of the two bodies, say Body A; the force exerted by Body A on Body B is called the "action", and the force exerted by Body B on Body A is called the "reaction". This law is sometimes referred to as the action-reaction law, with F_A called the "action" and F_B the "reaction". In other situations the magnitude and directions of the forces are determined jointly by both bodies and it isn't necessary to identify one force as the "action" and the other as the "reaction". The action and the reaction are simultaneous, and it does not matter which is called the action and which is called reaction; both forces are part of a single interaction, and neither force exists without the other.

The two forces in Newton's third law are of the same type (e.g., if the road exerts a forward frictional force on an accelerating car's tires, then it is also a frictional force that Newton's third law predicts for the tires pushing backward on the road).

From a conceptual standpoint, Newton's third law is seen when a person walks: they push against the floor, and the floor pushes against the person. Similarly, the tires of a car push against the road while the road pushes back on the tires—the tires and road simultaneously push against each other. In swimming, a person interacts with the water, pushing the water backward, while the water simultaneously pushes the person forward—both the person and the water push against each other. The reaction forces account for the motion in these examples. These forces depend on friction; a person or car on ice, for example, may be unable to exert the action force to produce the

Newton used the third law to derive the law of conservation of momentum; from a deeper perspective, however, conservation of momentum is the more fundamental idea (derived via Noether's theorem from Galilean invariance), and holds in cases where Newton's third law appears to fail, for instance when force fields as well as particles carry momentum, and in quantum mechanics.

Importance and Range of Validity

Newton's laws were verified by experiment and observation for over 200 years, and they are excel-lent approximations at the scales and speeds of everyday life. Newton's laws of motion, together with his law of universal gravitation and the mathematical techniques of calculus, provided for the first time a unified quantitative explanation for a wide range of physical phenomena.

These three laws hold to a good approximation for macroscopic objects under everyday conditions. However, Newton's laws (combined with universal gravitation and classical electrodynamics) are inappropriate for use in certain circumstances, most notably at very small scales, at very high speeds, or in very strong gravitational fields. Therefore, the laws cannot be used to explain phenomena such as conduction of electricity in a semiconductor, optical properties of substances, errors in non-relativistically corrected GPS systems and superconductivity. Explanation of these phenomena requires more sophisticated physical theories, including general relativity and quantum field theory.

In quantum mechanics, concepts such as force, momentum, and position are defined by linear operators that operate on the quantum state; at speeds that are much lower than the speed of light, Newton's laws are just as exact for these operators as they are for classical objects. At speeds comparable to the speed of light, the second law holds in the original form $F = dp/dt$, where F and p are four-vectors.

Relationship to the Conservation Laws

In modern physics, the laws of conservation of momentum, energy, and angular momentum are of more general validity than Newton's laws, since they apply to both light and matter, and to both classical and non-classical physics.

This can be stated simply, "Momentum, energy and angular momentum cannot be created or destroyed." Because force is the time derivative of momentum, the concept of force is redundant and subordinate to the conservation of momentum, and is not used in fundamental theories (e.g., quantum mechanics, quantum electrodynamics, general relativity, etc.). The standard model explains in detail how the three fundamental forces known as gauge forces originate out of exchange by virtual particles. Other forces, such as gravity and fermionic degeneracy pressure, also arise from the momen-tum conservation. Indeed, the conservation of 4-momentum in inertial motion via curved spacetime results in what we call gravitational force in general relativity theory. The application of the space derivative (which is a momentum operator in quantum mechanics) to the overlapping wave functions of a pair of fermions (particles with half-integer spin) results in shifts of maxima of compound wavefunction away from each other, which is observable as the "repulsion" of the fermions needed reaction force.

Newton stated the third law within a world-view that assumed instantaneous action at a distance between material particles. However, he was prepared for philosophical criticism of this action at a distance, and it was in this context that he stated the famous phrase "I feign no hypotheses". In modern physics, action at a distance has been completely eliminated, except for subtle effects involving quantum entanglement. (In particular, this refers to Bell's theorem–that no local model can reproduce the predictions of quantum theory.) Despite only being an approximation, in modern engineering and all practical applications involving the motion of vehicles and satellites, the concept of action at a distance is used extensively.

The discovery of the second law of thermodynamics by Carnot in the 19th century showed that not every physical quantity is conserved over time, thus disproving the validity of inducing the opposite metaphysical view from Newton's laws. Hence, a "steady-state" worldview based solely on Newton's laws and the conservation laws does not take entropy into account.

References

- Taylor, John R. (2005). Classical Mechanics. Sausalito, Calif.: Univ. Science Books. Pp. 133–138. ISBN 1-891389-22-X
- For example, P. K. Srivastava (2004). Mechanics. New Age International Pub. (P) Limited. P. 94. ISBN 9788122411126. Retrieved 2018-11-20
- Kobayashi, Yukio (2008). "Remarks on viewing situation in a rotating frame". European Journal of Physics. 29 (3): 599–606. Bibcode:2008ejph...29..599K. Doi:10.1088/0143-0807/29/3/019
- Colwell, Catharine H. "A Derivation of the Formulas for Centripetal Acceleration". Physicslab. Retrieved 31 July 2011
- Russelkl C Hibbeler (2009). "Equations of Motion: Normal and tangential coordinates". Engineering Mechanics: Dynamics (12 ed.). Prentice Hall. P. 131. ISBN 978-0-13-607791-6
- Centrifugal force". Encyclopædia Britannica. 17 August 2016. Retrieved 20 April 2017
- Https://www.britannica.com/science/friction
- Friction, science: britannica.com, Retrieved 15 February, 2019
- Van Berkel, Klaas (2013), Isaac Beeckman on Matter and Motion: Mechanical Philosophy in the Making, Johns Hopkins University Press, pp. 105–110, ISBN 9781421409368

4
Rotational Motion

The motion of a rigid body that occurs in a circular orbit about an axis with a common angular velocity is known as rotational motion. Some of its aspects are moment of inertia, angular momentum, torque, rolling, etc. This chapter has been carefully written to provide an easy understanding of these various aspects related to rotational motion.

Rotational motion is the motion of a rigid body which takes place in such a way that all of its particles move in circles about an axis with a common angular velocity; also, the rotation of a particle about a fixed point in space. Rotational motion is illustrated by (1) the fixed speed of rotation of the Earth about its axis; (2) the varying speed of rotation of the flywheel of a sewing machine; (3) the rotation of a satellite about a planet; (4) the motion of an ion in a cyclotron; and (5) the motion of a pendulum. Circular motion is a rotational motion in which each particle of the rotating body moves in a circular path about an axis. Such motion is exhibited by the first and second examples.

The speed of rotation, or angular velocity, remains constant in uniform circular motion. In this case, the angular displacement Θ experienced by the particle or rotating body in a time t is $\Theta = \omega t$, where ω is the constant angular velocity.

A special case of circular motion occurs when the rotating body moves with constant angular acceleration. If a body is moving in a circle with an angular acceleration of α radians/s^2, and if at a certain instant it has an angular velocity ω_0, then at a time t seconds later, the angular velocity may be expressed as $\omega = \omega_0 + \alpha t$, and the angular displacement as $\Theta = \omega_0 t + ½\alpha t^2$.

A rotating body possesses kinetic energy of rotation which may be expressed as $T_{rot} = ½I\omega^2$, where ω is the magnitude of the angular velocity of the rotating body and I is the moment of inertia, which is a measure of the opposition of the body to angular acceleration. The moment of inertia of a body depends on the mass of a body and the distribution of the mass relative to the axis of rotation. For example, the moment of inertia of a solid cylinder of mass M and radius R about its axis of symmetry is $½MR^2$.

The action of a torque L is to produce an angular acceleration α according to the equation below, where $I\omega$, the product,

$$L - Ia = \frac{dw}{dt} = \frac{d}{dt}(I\omega)$$

of moment of inertia and angular velocity, is called the angular momentum of the rotating body. This equation points out that the angular momentum $I\omega$ of a rotating body, and hence its angular velocity ω, remains constant unless the rotating body is acted upon by a torque. Both L and $I\omega$ may be represented by vectors.

It is readily shown that the work done by the torque L acting through an angle Θ on a rotating body originally at rest is exactly equal to the kinetic energy of rotation.

Moment of Inertia

The moment of inertia, otherwise known as the angular mass or rotational inertia, of a rigid body is a quantity that determines the torque needed for a desired angular acceleration about a rotational axis; similar to how mass determines the force needed for a desired acceleration. It depends on the body's mass distribution and the axis chosen, with larger moments requiring more torque to change the body's rotation rate. It is an extensive (additive) property: for a point mass the moment of inertia is just the mass times the square of the perpendicular distance to the rotation axis. The moment of inertia of a rigid composite system is the sum of the moments of inertia of its component subsystems (all taken about the same axis). Its simplest definition is the second moment of mass with respect to distance from an axis. For bodies constrained to rotate in a plane, only their moment of inertia about an axis perpendicular to the plane, a scalar value, matters. For bodies free to rotate in three dimensions, their moments can be described by a symmetric 3 × 3 matrix, with a set of mutually perpendicular principal axes for which this matrix is diagonal and torques around the axes act independently of each other.

When a body is free to rotate around an axis, torque must be applied to change its angular momentum. The amount of torque needed to cause any given angular acceleration (the rate of change in angular velocity) is proportional to the moment of inertia of the body. Moment of inertia may be expressed in units of kilogram meter squared ($kg{\cdot}m^2$) in SI units and pound-foot-second squared ($lbf{\cdot}ft{\cdot}s^2$) in imperial or US units.

Moment of inertia plays the role in rotational kinetics that mass (inertia) plays in linear kinetics - both characterize the resistance of a body to changes in its motion. The moment of inertia depends on how mass is distributed around an axis of rotation, and will vary depending on the chosen axis. For a point-like mass, the moment of inertia about some axis is given by mr^2, where r is the distance of the point from the axis, and m is the mass. For an extended rigid body, the moment of inertia is just the sum of all the small pieces of mass multiplied by the square of their distances from the axis in question. For an extended body of a regular shape and uniform density, this summation sometimes produces a simple expression that depends on the dimensions, shape and total mass of the object.

In 1673 Christiaan Huygens introduced this parameter in his study of the oscillation of a body hanging from a pivot, known as a compound pendulum. The term moment of inertia was introduced by Leonhard Euler in his book Theoria motus corporum solidorum seu rigidorum in 1765, and it is incorporated into Euler's second law.

The natural frequency of oscillation of a compound pendulum is obtained from the ratio of the torque imposed by gravity on the mass of the pendulum to the resistance to acceleration defined by the moment of inertia. Comparison of this natural frequency to that of a simple pendulum consisting of a single point of mass provides a mathematical formulation for moment of inertia of an extended body.

Moment of inertia also appears in momentum, kinetic energy, and in Newton's laws of motion for a rigid body as a physical parameter that combines its shape and mass. There is an interesting difference in the way moment of inertia appears in planar and spatial movement. Planar movement has a single scalar that defines the moment of inertia, while for spatial movement the same calculations yield a 3 × 3 matrix of moments of inertia, called the inertia matrix or inertia tensor.

The moment of inertia of a rotating flywheel is used in a machine to resist variations in applied torque to smooth its rotational output. The moment of inertia of an airplane about its longitudinal, horizontal and vertical axis determines how steering forces on the control surfaces of its wings, elevators and tail affect the plane in roll, pitch and yaw.

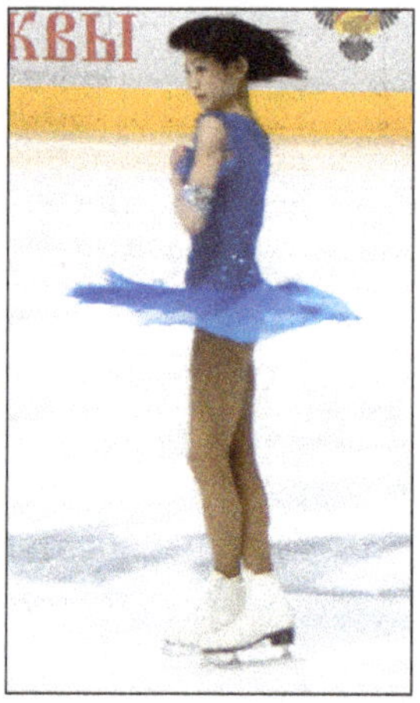

Spinning figure skaters can reduce their moment of inertia by pulling in their arms, allowing them to spin faster due to conservation of angular momentum.

Moment of inertia I is defined as the ratio of the net angular momentum L of a system to its angular velocity ω around a principal axis, that is

$$I = \frac{L}{\omega}.$$

If the angular momentum of a system is constant, then as the moment of inertia gets smaller, the angular velocity must increase. This occurs when spinning figure skaters pull in their outstretched arms or divers curl their bodies into a tuck position during a dive, to spin faster.

If the shape of the body does not change, then its moment of inertia appears in Newton's law of motion as the ratio of an applied torque τ on a body to the angular acceleration α around a principal axis, that is

$$\tau = I\alpha.$$

For a simple pendulum, this definition yields a formula for the moment of inertia I in terms of the mass m of the pendulum and its distance r from the pivot point as,

$$I = mr^2.$$

Thus, moment of inertia depends on both the mass m of a body and its geometry, or shape, as defined by the distance r to the axis of rotation.

This simple formula generalizes to define moment of inertia for an arbitrarily shaped body as the sum of all the elemental point masses dm each multiplied by the square of its perpendicular distance r to an axis k.

In general, given an object of mass m, an effective radius k can be defined for an axis through its center of mass, with such a value that its moment of inertia is,

$$I = mk^2,$$

where k is known as the radius of gyration.

Examples:

Simple Pendulum

Moment of inertia can be measured using a simple pendulum, because it is the resistance to the rotation caused by gravity. Mathematically, the moment of inertia of the pendulum is the ratio of the torque due to gravity about the pivot of a pendulum to its angular acceleration about that pivot point. For a simple pendulum this is found to be the product of the mass of the particle m with the square of its distance r to the pivot, that is

$$I = mr^2.$$

This can be shown as follows: The force of gravity on the mass of a simple pendulum generates a torque $\tau = \mathbf{r} \times \mathbf{F}$ around the axis perpendicular to the plane of the pendulum movement. Here $\mathbf{r}$ is the distance vector perpendicular to and from the force to the torque axis, and $\mathbf{F}$ is the net force on the mass. Associated with this torque is an angular acceleration, α, of the string and mass around this axis. Since the mass is constrained to a circle the tangential acceleration of the mass is $\mathbf{a} = \alpha \times \mathbf{r}$. Since $F = ma$ the torque equation becomes:

$$\begin{aligned}\tau &= \mathbf{r} \times \mathbf{F} = \mathbf{r} \times (m\alpha \times \mathbf{r}) \\ &= m((\mathbf{r} \cdot \mathbf{r})\alpha - (\mathbf{r} \cdot \alpha)\mathbf{r}) \\ &= mr^2\alpha = I\alpha\hat{\mathbf{k}},\end{aligned}$$

where $\hat{\mathbf{k}}$ is a unit vector perpendicular to the plane of the pendulum. (The second to last step uses the vector triple product expansion with the perpendicularity of α and $\mathbf{r}$.) The quantity $I = mr^2$ is the *moment of inertia* of this single mass around the pivot point.

The quantity $I = mr^2$ also appears in the angular momentum of a simple pendulum, which is calculated from the velocity $\mathbf{v} = \omega \times \mathbf{r}$ of the pendulum mass around the pivot, where ω is the angular velocity of the mass about the pivot point. This angular momentum is given by,

$$\begin{aligned}\mathbf{L} &= \mathbf{r} \times \mathbf{p} = \mathbf{r} \times (m\omega \times \mathbf{r}) \\ &= m((\mathbf{r} \cdot \mathbf{r})\omega - (\mathbf{r} \cdot \omega)\mathbf{r}) \\ &= mr^2\omega = I\omega\hat{\mathbf{k}},\end{aligned}$$

using a similar derivation to the previous equation.

Similarly, the kinetic energy of the pendulum mass is defined by the velocity of the pendulum around the pivot to yield,

$$E_K = \frac{1}{2}m\mathbf{v}\cdot\mathbf{v} = \frac{1}{2}\left(mr^2\right)\omega^2 = \frac{1}{2}I\omega^2.$$

This shows that the quantity $I = mr^2$ is how mass combines with the shape of a body to define rotational inertia. The moment of inertia of an arbitrarily shaped body is the sum of the values mr^2 for all of the elements of mass in the body.

Compound Pendulum

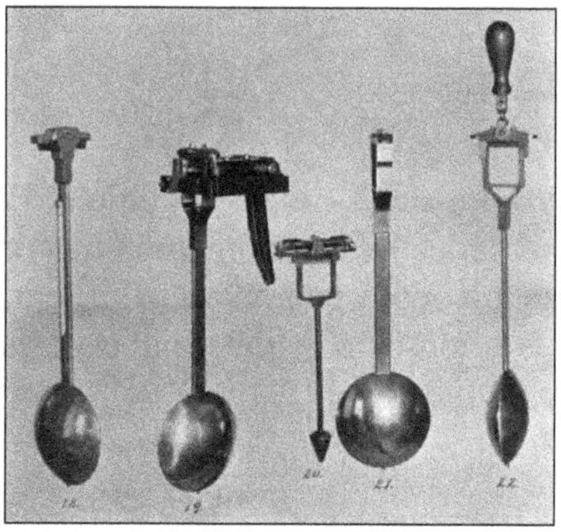

Pendulums used in Mendenhall gravimeter apparatus.
The portable gravimeter provided the most accurate relative measurements of the local gravitational field of the Earth.

A compound pendulum is a body formed from an assembly of particles of continuous shape that rotates rigidly around a pivot. Its moment of inertia is the sum of the moments of inertia of each of the particles that it is composed of.[:395–396:51–53] The natural frequency (ω_n) of a compound pendulum depends on its moment of inertia, I_P,

$$\omega_n = \sqrt{\frac{mgr}{I_P}},$$

where m is the mass of the object, g is local acceleration of gravity, and r is the distance from the pivot point to the center of mass of the object. Measuring this frequency of oscillation over small angular displacements provides an effective way of measuring moment of inertia of a body.

Thus, to determine the moment of inertia of the body, simply suspend it from a convenient pivot point P so that it swings freely in a plane perpendicular to the direction of the desired moment of inertia, then measure its natural frequency or period of oscillation (t), to obtain

$$I_P = \frac{mgr}{\omega_n^2} = \frac{mgrt^2}{4\pi^2},$$

where t is the period (duration) of oscillation (usually averaged over multiple periods).

The moment of inertia of the body about its center of mass, I_C, is then calculated using the parallel axis theorem to be,

$$I_C = I_P - mr^2,$$

where m is the mass of the body and r is the distance from the pivot point P to the center of mass C.

Moment of inertia of a body is often defined in terms of its *radius of gyration*, which is the radius of a ring of equal mass around the center of mass of a body that has the same moment of inertia. The radius of gyration k is calculated from the body's moment of inertia I_C and mass m as the lenght.

$$k = \sqrt{\frac{I_C}{m}}.$$

Center of Oscillation

A simple pendulum that has the same natural frequency as a compound pendulum defines the length L from the pivot to a point called the center of oscillation of the compound pendulum. This point also corresponds to the center of percussion. The length L is determined from the formula,

$$\omega_n = \sqrt{\frac{g}{L}} = \sqrt{\frac{mgr}{I_P}},$$

or

$$L = \frac{g}{\omega_n^2} = \frac{I_P}{mr}.$$

The seconds pendulum, which provides the "tick" and "tock" of a grandfather clock, takes one second to swing from side-to-side. This is a period of two seconds, or a natural frequency of π rad/s for the pendulum. In this case, the distance to the center of oscillation, L, can be computed to be,

$$L = \frac{g}{\omega_n^2} \approx \frac{9.81 \text{ m/s}^2}{(3.14 \text{ rad/s})^2} \approx 0.99 \text{ m}.$$

Notice that the distance to the center of oscillation of the seconds pendulum must be adjusted to accommodate different values for the local acceleration of gravity. Kater's pendulum is a compound pendulum that uses this property to measure the local acceleration of gravity, and is called a gravimeter.

Measuring Moment of Inertia

The moment of inertia of a complex system such as a vehicle or airplane around its vertical axis can be measured by suspending the system from three points to form a trifilar pendulum. A trifilar pendulum is a platform supported by three wires designed to oscillate in torsion around its vertical centroidal axis. The period of oscillation of the trifilar pendulum yields the moment of inertia of the system.

Motion in a Fixed Plane

Point Mass

The moment of inertia about an axis of a body is calculated by summing mr^2 for every particle in the body, where r is the perpendicular distance to the specified axis. To see how moment of inertia arises in the study of the movement of an extended body, it is convenient to consider a rigid assembly of point masses. (This equation can be used for axes that are not principal axes provided that it is understood that this does not fully describe the moment of inertia.)

Consider the kinetic energy of an assembly of N masses m_i that lie at the distances r_i from the pivot point P, which is the nearest point on the axis of rotation. It is the sum of the kinetic energy of the individual masses,

$$E_{\mathrm{K}} = \sum_{i=1}^{N} \frac{1}{2} m_i \mathbf{v}_i \cdot \mathbf{v}_i = \sum_{i=1}^{N} \frac{1}{2} m_i \left(\omega r_i\right)^2 = \frac{1}{2} \omega^2 \sum_{i=1}^{N} m_i r_i^2 .$$

This shows that the moment of inertia of the body is the sum of each of the mr^2 terms, that is

$$I_P = \sum_{i=1}^{N} m_i r_i^2 .$$

Thus, moment of inertia is a physical property that combines the mass and distribution of the particles around the rotation axis. Notice that rotation about different axes of the same body yield different moments of inertia.

The moment of inertia of a continuous body rotating about a specified axis is calculated in the same way, except with infinitely many point particles. Thus the limits of summation are removed, and the sum is written as follows:

$$I_P = \sum_{i} m_i r_i^2$$

Another expression replaces the summation with an integral,

$$I_P = \iiint_{Q} \rho(x,y,z) \left\| \mathbf{r} \right\|^2 \mathrm{d}V$$

Here, the function ρ gives the mass density at each point (x,y,z), $\mathbf{r}$ is a vector perpendicular to the axis of rotation and extending from a point on the rotation axis to a point (x,y,z) in the solid, and the integration is evaluated over the volume V of the body Q. The moment of inertia of a flat surface is similar with the mass density being replaced by its areal mass density with the integral evaluated over its area.

Note on second moment of area: The moment of inertia of a body moving in a plane and the second moment of area of a beam's cross-section are often confused. The moment of inertia of a body with the shape of the cross-section is the second moment of this area about the -axis perpendicular to the cross-section, weighted by its density. This is also called the *polar moment of the area*, and is the sum of the second moments about the x- and y-axes. The stresses in a beam are calculated using the second moment of the cross-sectional area around either the x-axis or y-axis depending on the load.

Examples

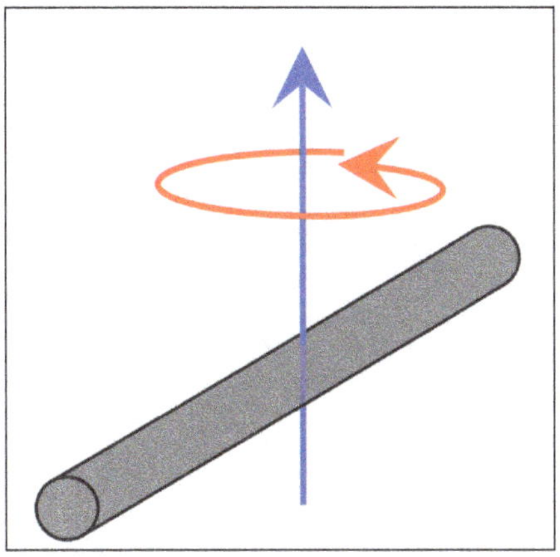

The moment of inertia of a compound pendulum constructed from a thin disc mounted at the end of a thin rod that oscillates around a pivot at the other end of the rod, begins with the calculation of the moment of inertia of the thin rod and thin disc about their respective centers of mass.

- The moment of inertia of a thin rod with constant cross-section s and density ρ and with length ℓ about a perpendicular axis through its center of mass is determined by integration. Align the x-axis with the rod and locate the origin its center of mass at the center of the rod, then

$$I_{C,\text{rod}} = \iiint_Q \rho x^2 \,\mathrm{d}V = \int_{-\frac{\ell}{2}}^{\frac{\ell}{2}} \rho x^2 s \,\mathrm{d}x = \rho s \frac{x^3}{3}\bigg|_{-\frac{\ell}{2}}^{\frac{\ell}{2}} = \frac{\rho s}{3}\left(\frac{\ell^3}{8} + \frac{\ell^3}{8}\right) = \frac{m\ell^2}{12},$$

  where $m = \rho s \ell$ is the mass of the rod.

- The moment of inertia of a thin disc of constant thickness s, radius R, and density ρ about an axis through its center and perpendicular to its face (parallel to its axis of rotational symmetry) is determined by integration. Align the z-axis with the axis of the disc and define a volume element as $\mathrm{d}V = sr\mathrm{d}r\mathrm{d}\theta$, then

$$I_{C,\text{disc}} = \iiint_Q \rho r^2 \,\mathrm{d}V = \int_0^{2\pi}\int_0^R \rho r^2 sr\,\mathrm{d}r\,\mathrm{d}\theta = 2\pi\rho s \frac{R^4}{4} = \frac{1}{2} mR^2,$$

  where $m = \pi R^2 \rho s$ is its mass.

- The moment of inertia of the compound pendulum is now obtained by adding the moment of inertia of the rod and the disc around the pivot point P as,

$$I_P = I_{C,\text{rod}} + M_{\text{rod}}\left(\frac{L}{2}\right)^2 + I_{C,\text{disc}} + M_{\text{disc}}(L+R)^2,$$

  where L is the length of the pendulum. Notice that the parallel axis theorem is used to shift the moment of inertia from the center of mass to the pivot point of the pendulum.

A list of moments of inertia formulas for standard body shapes provides a way to obtain the moment of inertia of a complex body as an assembly of simpler shaped bodies. The parallel axis theorem is used to shift the reference point of the individual bodies to the reference point of the assembly.

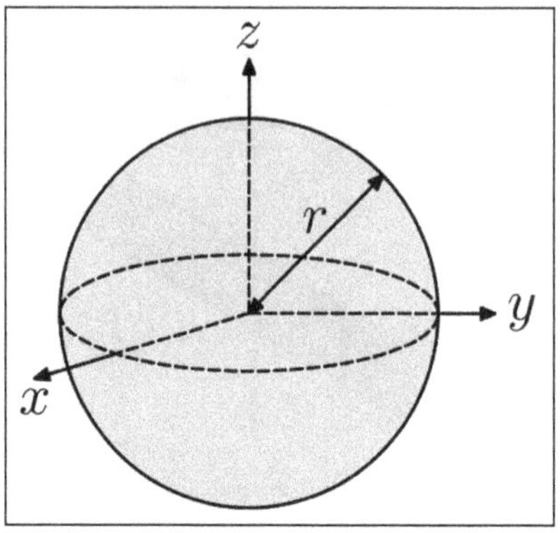

As one more example, consider the moment of inertia of a solid sphere of constant density about an axis through its center of mass. This is determined by summing the moments of inertia of the thin discs that form the sphere. If the surface of the ball is defined by the equation,

$$x^2 + y^2 + z^2 = R^2,$$

then the radius r of the disc at the cross-section z along the z-axis is,

$$r(z)^2 = x^2 + y^2 = R^2 - z^2.$$

Therefore, the moment of inertia of the ball is the sum of the moments of inertia of the discs along the z-axis,

$$\begin{aligned} I_{C,\text{ball}} &= \int_{-R}^{R} \frac{\pi\rho}{2} r(z)^4 \,\mathrm{d}z = \int_{-R}^{R} \frac{\pi\rho}{2} \left(R^2 - z^2\right)^2 \mathrm{d}z \\ &= \frac{\pi\rho}{2} \left(R^4 z - \frac{2}{3} R^2 z^3 + \frac{1}{5} z^5 \right)\Bigg|_{-R}^{R} \\ &= \pi\rho \left(1 - \frac{2}{3} + \frac{1}{5} \right) R^5 \\ &= \frac{2}{5} m R^2, \end{aligned}$$

where $m = \frac{4}{3}\pi R^3 \rho$ is the mass of the sphere.

Rigid Body

If a mechanical system is constrained to move parallel to a fixed plane, then the rotation of a body in the system occurs around an axis $\hat{\mathbf{k}}$ perpendicular to this plane. In this case, the moment of inertia of the mass in this system is a scalar known as the polar moment of inertia. The definition of the polar moment of inertia can be obtained by considering momentum, kinetic energy and Newton's laws for the planar movement of a rigid system of particles.

If a system of n particles, $P_i, i=1,\ldots,n$, are assembled into a rigid body, then the momentum of the system can be written in terms of positions relative to a reference point $\mathbf{R}$, and absolute velocities $\mathbf{v}_i$,

$$\Delta\mathbf{r}_i = \mathbf{r}_i - \mathbf{R},$$
$$\mathbf{v}_i = \boldsymbol{\omega}\times(\mathbf{r}_i - \mathbf{R}) + \mathbf{V} = \boldsymbol{\omega}\times\Delta\mathbf{r}_i + \mathbf{V},$$

where $\boldsymbol{\omega}$ is the angular velocity of the system and $\mathbf{V}$ is the velocity of $\mathbf{R}$.

For planar movement the angular velocity vector is directed along the unit vector $\mathbf{k}$ which is perpendicular to the plane of movement. Introduce the unit vectors $\mathbf{e}_i$ from the reference point $\mathbf{R}$ to a point $\mathbf{r}_i$, and the unit vector $\hat{\mathbf{t}}_i = \hat{\mathbf{k}}\times\hat{\mathbf{e}}_i$, so

$$\hat{\mathbf{e}}_i = \frac{\Delta\mathbf{r}_i}{\Delta r_i}, \quad \hat{\mathbf{k}} = \frac{\boldsymbol{\omega}}{\omega}, \quad \hat{\mathbf{t}}_i = \hat{\mathbf{k}}\times\hat{\mathbf{e}}_i,$$
$$\mathbf{v}_i = \boldsymbol{\omega}\times\Delta\mathbf{r}_i + \mathbf{V} = \omega\hat{\mathbf{k}}\times\Delta r_i\hat{\mathbf{e}}_i + \mathbf{V} = \omega\Delta r_i\hat{\mathbf{t}}_i + \mathbf{V}$$

This defines the relative position vector and the velocity vector for the rigid system of the particles moving in a plane.

Note on the cross product: When a body moves parallel to a ground plane, the trajectories of all the points in the body lie in planes parallel to this ground plane. This means that any rotation that the body undergoes must be around an axis perpendicular to this plane. Planar movement is often presented as projected onto this ground plane so that the axis of rotation appears as a point. In this case, the angular velocity and angular acceleration of the body are scalars and the fact that they are vectors along the rotation axis is ignored. This is usually preferred for introductions to the topic. But in the case of moment of inertia, the combination of mass and geometry benefits from the geometric properties of the cross product. For this reason, in this section on planar movement the angular velocity and accelerations of the body are vectors perpendicular to the ground plane, and the cross product operations are the same as used for the study of spatial rigid body movement.

Angular Momentum

The angular momentum vector for the planar movement of a rigid system of particles is given by,

$$\mathbf{L} = \sum_{i=1}^{n} m_i\Delta\mathbf{r}_i\times\mathbf{v}_i$$
$$= \sum_{i=1}^{n} m_i\,\Delta r_i\hat{\mathbf{e}}_i\times(\omega\Delta r_i\hat{\mathbf{t}}_i + \mathbf{V})$$
$$= \left(\sum_{i=1}^{n} m_i\,\Delta r_i^2\right)\omega\hat{\mathbf{k}} + \left(\sum_{i=1}^{n} m_i\,\Delta r_i\hat{\mathbf{e}}_i\right)\times\mathbf{V}.$$

Use the center of mass $\mathbf{C}$ as the reference point so,

$$\Delta r_i\hat{\mathbf{e}}_i = \mathbf{r}_i - \mathbf{C},$$
$$\sum_{i=1}^{n} m_i\,\Delta r_i\hat{\mathbf{e}}_i = 0,$$

and define the moment of inertia relative to the center of mass I_C as,

$$I_C = \sum_i m_i \Delta r_i^2,$$

then the equation for angular momentum simplifies to,

$$\mathbf{L} = I_C \omega \hat{\mathbf{k}}.$$

The moment of inertia about an axis perpendicular to the movement of the rigid system and through the center of mass is known as the *polar moment of inertia*. Specifically, it is the second moment of mass with respect to the orthogonal distance from an axis (or pole).

For a given amount of angular momentum, a decrease in the moment of inertia results in an increase in the angular velocity. Figure skaters can change their moment of inertia by pulling in their arms. Thus, the angular velocity achieved by a skater with outstretched arms results in a greater angular velocity when the arms are pulled in, because of the reduced moment of inertia. A figure skater is not, however, a rigid body.

Kinetic Energy

The kinetic energy of a rigid system of particles moving in the plane is given by,

$$\begin{aligned} E_K &= \frac{1}{2}\sum_{i=1}^{n} m_i \mathbf{v}_i \cdot \mathbf{v}_i, \\ &= \frac{1}{2}\sum_{i=1}^{n} m_i \left(\omega \Delta r_i \hat{\mathbf{t}}_i + \mathbf{V}\right)\cdot\left(\omega \Delta r_i \hat{\mathbf{t}}_i + \mathbf{V}\right), \\ &= \frac{1}{2}\omega^2 \left(\sum_{i=1}^{n} m_i \Delta r_i^2 \hat{\mathbf{t}}_i \cdot \hat{\mathbf{t}}_i\right) + \omega \mathbf{V}\cdot\left(\sum_{i=1}^{n} m_i \Delta r_i \hat{\mathbf{t}}_i\right) + \frac{1}{2}\left(\sum_{i=1}^{n} m_i\right)\mathbf{V}\cdot\mathbf{V} \end{aligned}$$

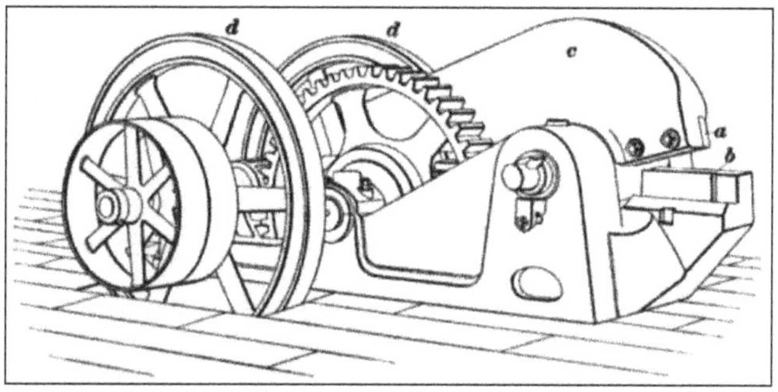

This rotary shear uses the moment of inertia of two flywheels to store kinetic energy which when released is used to cut metal stock.

Let the reference point be the center of mass **C** of the system so the second term becomes zero, and introduce the moment of inertia I_C so the kinetic energy is given by,

$$E_K = \frac{1}{2} I_C \omega^2 + \frac{1}{2} M\mathbf{V}\cdot\mathbf{V}.$$

The moment of inertia I_C is the *polar moment of inertia* of the body.

Newton's Laws

A John Deere tractor with the spoked flywheel on the engine. The large moment of inertia of the flywheel smooths the operation of the tractor.

Newton's laws for a rigid system of n particles, $P_i, i=1,\dots,n$, can be written in terms of a resultant force and torque at a reference point $\mathbf{R}$, to yield,

$$\mathbf{F} = \sum_{i=1}^{n} m_i \mathbf{A}_i,$$

$$\tau = \sum_{i=1}^{n} \Delta\mathbf{r}_i \times m_i \mathbf{A}_i,$$

where $\mathbf{r}_i$ denotes the trajectory of each particle.

The kinematics of a rigid body yields the formula for the acceleration of the particle P_i in terms of the position $\mathbf{R}$ and acceleration $\mathbf{A}$ of the reference particle as well as the angular velocity vector **ù** and angular acceleration vector **á** of the rigid system of particles as,

$$\mathbf{A}_i = \boldsymbol{\alpha} \times \Delta\mathbf{r}_i + \boldsymbol{\omega} \times \boldsymbol{\omega} \times \Delta\mathbf{r}_i + \mathbf{A}.$$

For systems that are constrained to planar movement, the angular velocity and angular acceleration vectors are directed along $\hat{\mathbf{k}}$ perpendicular to the plane of movement, which simplifies this acceleration equation. In this case, the acceleration vectors can be simplified by introducing the unit vectors $\hat{\mathbf{e}}_i$ from the reference point $\mathbf{R}$ to a point $\mathbf{r}_i$ and the unit vectors $\hat{\mathbf{t}}_i = \hat{\mathbf{k}} \times \hat{\mathbf{e}}_i$, so

$$\begin{aligned}\mathbf{A}_i &= \alpha\hat{\mathbf{k}} \times \Delta r_i \hat{\mathbf{e}}_i - \omega\hat{\mathbf{k}} \times \omega\hat{\mathbf{k}} \times \Delta r_i \hat{\mathbf{e}}_i + \mathbf{A} \\ &= \alpha \Delta r_i \hat{\mathbf{t}}_i - \omega^2 \Delta r_i \hat{\mathbf{e}}_i + \mathbf{A}.\end{aligned}$$

This yields the resultant torque on the system as,

$$\begin{aligned}\tau &= \sum_{i=1}^{n} m_i \Delta r_i \hat{\mathbf{e}}_i \times \left(\alpha \Delta r_i \hat{\mathbf{t}}_i - \omega^2 \Delta r_i \hat{\mathbf{e}}_i + \mathbf{A}\right) \\ &= \left(\sum_{i=1}^{n} m_i \Delta r_i^2\right)\alpha\hat{\mathbf{k}} + \left(\sum_{i=1}^{n} m_i \Delta r_i \hat{\mathbf{e}}_i\right) \times \mathbf{A},\end{aligned}$$

where $\hat{\mathbf{e}}_i \times \hat{\mathbf{e}}_i = \mathbf{0}$, and $\hat{\mathbf{e}}_i \times \hat{\mathbf{t}}_i = \hat{\mathbf{k}}$ is the unit vector perpendicular to the plane for all of the particles P_i.

Use the center of mass $\mathbf{C}$ as the reference point and define the moment of inertia relative to the center of mass $I_{\mathbf{C}}$, then the equation for the resultant torque simplifies to,

$$\tau = I_{\mathbf{C}}\alpha\hat{\mathbf{k}}$$

Motion in Space of a Rigid Body and the Inertia Matrix

The scalar moments of inertia appear as elements in a matrix when a system of particles is assembled into a rigid body that moves in three-dimensional space. This inertia matrix appears in the calculation of the angular momentum, kinetic energy and resultant torque of the rigid system of particles.

Let the system of n particles, $P_i, i = 1,\ldots,n$ be located at the coordinates $\mathbf{r}_i$ with velocities $\mathbf{v}_i$ relative to a fixed reference frame. For a (possibly moving) reference point $\mathbf{R}$, the relative positions are,

$$\Delta\mathbf{r}_i = \mathbf{r}_i - \mathbf{R}$$

and the (absolute) velocities are,

$$\mathbf{v}_i = \omega \times \Delta\mathbf{r}_i + \mathbf{V}_{\mathbf{R}}$$

where ω is the angular velocity of the system, and $\mathbf{V}_{\mathbf{R}}$ is the velocity of $\mathbf{R}$.

Angular Momentum

Note that the cross product can be equivalently written as matrix multiplication by combining the first operand and the operator into a, skew-symmetric, matrix, $[\mathbf{b}]$, constructed from the components of $\mathbf{b} = (b_x, b_y, b_z)$:

$$\mathbf{b} \times \mathbf{y} \equiv [\mathbf{b}]\mathbf{y}$$

$$[\mathbf{b}] \equiv \begin{bmatrix} 0 & -b_z & b_y \\ b_z & 0 & -b_x \\ -b_y & b_x & 0 \end{bmatrix}.$$

The inertia matrix is constructed by considering the angular momentum, with the reference point $\mathbf{R}$ of the body chosen to be the center of mass $\mathbf{C}$:

$$\begin{aligned} \mathbf{L} &= \sum_{i=1}^{n} m_i \Delta\mathbf{r}_i \times \mathbf{v}_i \\ &= \sum_{i=1}^{n} m_i \Delta\mathbf{r}_i \times (\omega \times \Delta\mathbf{r}_i + \mathbf{V}_{\mathbf{R}}) \\ &= \left(-\sum_{i=1}^{n} m_i \Delta\mathbf{r}_i \times (\Delta\mathbf{r}_i \times \omega)\right) + \left(\sum_{i=1}^{n} m_i \Delta\mathbf{r}_i \times \mathbf{V}_{\mathbf{R}}\right), \end{aligned}$$

where the terms containing $\mathbf{V}_\mathbf{R}$ $(=\mathbf{C})$ sum to zero by the definition of center of mass.

Then, the skew-symmetric matrix $[\Delta\mathbf{r}_i]$ obtained from the relative position vector $\Delta\mathbf{r}_i = \mathbf{r}_i - \mathbf{C}$, can be used to define,

$$\mathbf{L} = \left(-\sum_{i=1}^{n} m_i [\Delta\mathbf{r}_i]^2 \right) \boldsymbol{\omega} = \mathbf{I}_\mathrm{C}\boldsymbol{\omega},$$

where $\mathbf{I}_\mathrm{C}$ defined by,

$$\mathbf{I}_\mathrm{C} = -\sum_{i=1}^{n} m_i [\Delta\mathbf{r}_i]^2,$$

is the symmetric inertia matrix of the rigid system of particles measured relative to the center of mass $\mathbf{C}$.

Kinetic Energy

The kinetic energy of a rigid system of particles can be formulated in terms of the center of mass and a matrix of mass moments of inertia of the system. Let the system of n particles $P_i, i = 1,\ldots,n$ be located at the coordinates $\mathbf{r}_i$ with velocities $\mathbf{v}_i$, then the kinetic energy is,

$$E_\mathrm{K} = \frac{1}{2}\sum_{i=1}^{n} m_i \mathbf{v}_i \cdot \mathbf{v}_i = \frac{1}{2}\sum_{i=1}^{n} m_i (\boldsymbol{\omega}\times\Delta\mathbf{r}_i + \mathbf{V}_\mathrm{C})\cdot(\boldsymbol{\omega}\times\Delta\mathbf{r}_i + \mathbf{V}_\mathrm{C})$$

where $\Delta\mathbf{r}_i = \mathbf{r}_i - \mathbf{C}$ is the position vector of a particle relative to the center of mass.

This equation expands to yield three terms,

$$E_\mathrm{K} = \frac{1}{2}\left(\sum_{i=1}^{n} m_i (\boldsymbol{\omega}\times\Delta\mathbf{r}_i)\cdot(\boldsymbol{\omega}\times\Delta\mathbf{r}_i)\right) + \left(\sum_{i=1}^{n} m_i \mathbf{V}_\mathrm{C}\cdot(\boldsymbol{\omega}\times\Delta\mathbf{r}_i)\right) + \frac{1}{2}\left(\sum_{i=1}^{n} m_i \mathbf{V}_\mathrm{C}\cdot\mathbf{V}_\mathrm{C}\right).$$

The second term in this equation is zero because $\mathbf{C}$ is the center of mass. Introduce the skew-symmetric matrix $[\Delta\mathbf{r}_i]$ so the kinetic energy becomes,

$$\begin{aligned} E_\mathrm{K} &= \frac{1}{2}\left(\sum_{i=1}^{n} m_i ([\Delta\mathbf{r}_i]\boldsymbol{\omega})\cdot([\Delta\mathbf{r}_i]\boldsymbol{\omega})\right) + \frac{1}{2}\left(\sum_{i=1}^{n} m_i\right)\mathbf{V}_\mathrm{C}\cdot\mathbf{V}_\mathrm{C} \\ &= \frac{1}{2}\left(\sum_{i=1}^{n} m_i \left(\boldsymbol{\omega}^\mathrm{T}[\Delta\mathbf{r}_i]^\mathrm{T}[\Delta\mathbf{r}_i]\boldsymbol{\omega}\right)\right) + \frac{1}{2}\left(\sum_{i=1}^{n} m_i\right)\mathbf{V}_\mathrm{C}\cdot\mathbf{V}_\mathrm{C} \\ &= \frac{1}{2}\boldsymbol{\omega}\cdot\left(-\sum_{i=1}^{n} m_i [\Delta\mathbf{r}_i]^2\right)\boldsymbol{\omega} + \frac{1}{2}\left(\sum_{i=1}^{n} m_i\right)\mathbf{V}_\mathrm{C}\cdot\mathbf{V}_\mathrm{C}. \end{aligned}$$

Thus, the kinetic energy of the rigid system of particles is given by,

$$E_\mathrm{K} = \frac{1}{2}\boldsymbol{\omega}\cdot\mathbf{I}_\mathrm{C}\boldsymbol{\omega} + \frac{1}{2}M\mathbf{V}_\mathrm{C}^2.$$

where $\mathbf{I}_\mathrm{C}$ is the inertia matrix relative to the center of mass and M is the total mass.

Resultant Torque

The inertia matrix appears in the application of Newton's second law to a rigid assembly of particles. The resultant torque on this system is,

$$\tau = \sum_{i=1}^{n} (\mathbf{r}_i - \mathbf{R}) \times m_i \mathbf{a}_i,$$

where $\mathbf{a}_i$ is the acceleration of the particle P_i. The kinematics of a rigid body yields the formula for the acceleration of the particle P_i in terms of the position $\mathbf{R}$ and acceleration $\mathbf{A}_\mathbf{R}$ of the reference point, as well as the angular velocity vector ω and angular acceleration vector α of the rigid system as,

$$\mathbf{a}_i = \alpha \times (\mathbf{r}_i - \mathbf{R}) + \omega \times \omega \times (\mathbf{r}_i - \mathbf{R}) + \mathbf{A}_\mathbf{R}.$$

Use the center of mass $\mathbf{C}$ as the reference point, and introduce the skew-symmetric matrix $[\Delta\mathbf{r}_i] = [\mathbf{r}_i - \mathbf{C}]$ to represent the cross product $(\mathbf{r}_i - \mathbf{C})\times$, to obtain,

$$\tau = \left(-\sum_{i=1}^{n} m_i [\Delta\mathbf{r}_i]^2 \right) \alpha + \omega \times \left(-\sum_{i=1}^{n} m_i [\Delta\mathbf{r}_i]^2 \right) \omega$$

The calculation uses the identity,

$$\Delta\mathbf{r}_i \times (\omega \times (\omega \times \Delta\mathbf{r}_i)) + \omega \times ((\omega \times \Delta\mathbf{r}_i) \times \Delta\mathbf{r}_i) = 0,$$

obtained from the Jacobi identity for the triple cross product:

Thus, the resultant torque on the rigid system of particles is given by,

$$\tau = \mathbf{I}_\mathbf{C} \alpha + \omega \times \mathbf{I}_\mathbf{C} \omega,$$

where $\mathbf{I}_\mathbf{C}$ is the inertia matrix relative to the center of mass.

Parallel Axis Theorem

The inertia matrix of a body depends on the choice of the reference point. There is a useful relationship between the inertia matrix relative to the center of mass $\mathbf{C}$ and the inertia matrix relative to another point $\mathbf{R}$. This relationship is called the parallel axis theorem.

Consider the inertia matrix $\mathbf{I}_\mathbf{R}$ obtained for a rigid system of particles measured relative to a reference point $\mathbf{R}$, given by

$$\mathbf{I}_\mathbf{R} = -\sum_{i=1}^{n} m_i [\mathbf{r}_i - \mathbf{R}]^2.$$

Let $\mathbf{C}$ be the center of mass of the rigid system, then

$$\mathbf{R} = (\mathbf{R} - \mathbf{C}) + \mathbf{C} = \mathbf{d} + \mathbf{C},$$

where $\mathbf{d}$ is the vector from the center of mass $\mathbf{C}$ to the reference point $\mathbf{R}$. Use this equation to compute the inertia matrix,

$$\mathbf{I_R} = -\sum_{i=1}^{n} m_i[\mathbf{r}_i - (\mathbf{C}+\mathbf{d})]^2 = -\sum_{i=1}^{n} m_i[(\mathbf{r}_i - \mathbf{C}) - \mathbf{d}]^2$$

Distribute over the cross product to obtain,

$$\mathbf{I_R} = -\left(\sum_{i=1}^{n} m_i[\mathbf{r}_i - \mathbf{C}]^2\right) + \left(\sum_{i=1}^{n} m_i[\mathbf{r}_i - \mathbf{C}]\right)[\mathbf{d}] + [\mathbf{d}]\left(\sum_{i=1}^{n} m_i[\mathbf{r}_i - \mathbf{C}]\right) - \left(\sum_{i=1}^{n} m_i\right)[\mathbf{d}]^2$$

The first term is the inertia matrix $\mathbf{I_C}$ relative to the center of mass. The second and third terms are zero by definition of the center of mass $\mathbf{C}$. And the last term is the total mass of the system multiplied by the square of the skew-symmetric matrix $[\mathbf{d}]$ constructed from $\mathbf{d}$.

The result is the parallel axis theorem,

$$\mathbf{I_R} = \mathbf{I_C} - M[\mathbf{d}]^2,$$

where $\mathbf{d}$ is the vector from the center of mass $\mathbf{C}$ to the reference point $\mathbf{R}$.

Note on the minus sign: By using the skew symmetric matrix of position vectors relative to the reference point, the inertia matrix of each particle has the form $-m[\mathbf{r}]^2$, which is similar to the mr^2 that appears in planar movement. However, to make this to work out correctly a minus sign is needed. This minus sign can be absorbed into the term $m[\mathbf{r}]^{\mathsf{T}}[\mathbf{r}]$, if desired, by using the skew-symmetry property of $[\mathbf{r}]$.

Scalar Moment of Inertia in a Plane

The scalar moment of inertia, I_L, of a body about a specified axis whose direction is specified by the unit vector $\hat{\mathbf{k}}$ and passes through the body at a point $\mathbf{R}$ is as follows:

$$I_L = \hat{\mathbf{k}}\cdot\left(-\sum_{i=1}^{N} m_i[\Delta\mathbf{r}_i]^2\right)\hat{\mathbf{k}} = \hat{\mathbf{k}}\cdot\mathbf{I_R}\hat{\mathbf{k}} = \hat{\mathbf{k}}^{\mathsf{T}}\mathbf{I_R}\hat{\mathbf{k}},$$

where $\mathbf{I_R}$ is the moment of inertia matrix of the system relative to the reference point $\mathbf{R}$, and $[\Delta\mathbf{r}_i]$ is the skew symmetric matrix obtained from the vector $\Delta\mathbf{r}_i = \mathbf{r}_i - \mathbf{R}$.

This is derived as follows. Let a rigid assembly of n particles, $P_i, i = 1,\ldots,n$, have coordinates $\mathbf{r}_i$. Choose $\mathbf{R}$ as a reference point and compute the moment of inertia around a line L defined by the unit vector $\hat{\mathbf{k}}$ through the reference point $\mathbf{R}$, $\mathbf{L}(t) = \mathbf{R} + t\hat{\mathbf{k}}$. The perpendicular vector from this line to the particle P_i is obtained from $\Delta\mathbf{r}_i$ by removing the component that projects onto $\hat{\mathbf{k}}$.

$$\Delta\mathbf{r}_i^{\perp} = \Delta\mathbf{r}_i - (\hat{\mathbf{k}}\cdot\Delta\mathbf{r}_i)\hat{\mathbf{k}} = (\mathbf{E} - \hat{\mathbf{k}}\hat{\mathbf{k}}^{\mathsf{T}})\Delta\mathbf{r}_i,$$

where $\mathbf{E}$ is the identity matrix, so as to avoid confusion with the inertia matrix, and $\hat{\mathbf{k}}\hat{\mathbf{k}}^{\mathsf{T}}$ is the outer product matrix formed from the unit vector $\hat{\mathbf{k}}$ along the line L.

To relate this scalar moment of inertia to the inertia matrix of the body, introduce the skew-symmetric matrix $[\hat{\mathbf{k}}]$ such that $[\hat{\mathbf{k}}]\mathbf{y} = \hat{\mathbf{k}} \times \mathbf{y}$, then we have the identity,

$$-[\hat{\mathbf{k}}]^2 \equiv |\hat{\mathbf{k}}|^2 (\mathbf{E} - \hat{\mathbf{k}}\hat{\mathbf{k}}^\mathsf{T}) = \mathbf{E} - \hat{\mathbf{k}}\hat{\mathbf{k}}^\mathsf{T},$$

noting that $\hat{\mathbf{k}}$ is a unit vector.

The magnitude squared of the perpendicular vector is,

$$\begin{aligned} |\Delta\mathbf{r}_i^\perp|^2 &= \left(-[\hat{\mathbf{k}}]^2 \Delta\mathbf{r}_i\right)\cdot\left(-[\hat{\mathbf{k}}]^2 \Delta\mathbf{r}_i\right) \\ &= \left(\hat{\mathbf{k}} \times (\hat{\mathbf{k}} \times \Delta\mathbf{r}_i)\right)\cdot\left(\hat{\mathbf{k}} \times (\hat{\mathbf{k}} \times \Delta\mathbf{r}_i)\right) \end{aligned}$$

The simplification of this equation uses the triple scalar product identity,

$$\left(\hat{\mathbf{k}} \times (\hat{\mathbf{k}} \times \Delta\mathbf{r}_i)\right)\cdot\left(\hat{\mathbf{k}} \times (\hat{\mathbf{k}} \times \Delta\mathbf{r}_i)\right) \equiv \left(\left(\hat{\mathbf{k}} \times (\hat{\mathbf{k}} \times \Delta\mathbf{r}_i)\right) \times \hat{\mathbf{k}}\right)\cdot\left(\hat{\mathbf{k}} \times \Delta\mathbf{r}_i\right),$$

where the dot and the cross products have been interchanged. Exchanging products, and simplifying by noting that $\Delta\mathbf{r}_i$ and $\hat{\mathbf{k}}$ are orthogonal:

$$\begin{aligned} &\left(\hat{\mathbf{k}} \times (\hat{\mathbf{k}} \times \Delta\mathbf{r}_i)\right)\cdot\left(\hat{\mathbf{k}} \times (\hat{\mathbf{k}} \times \Delta\mathbf{r}_i)\right) \\ &= \left(\left(\hat{\mathbf{k}} \times (\hat{\mathbf{k}} \times \Delta\mathbf{r}_i)\right) \times \hat{\mathbf{k}}\right)\cdot\left(\hat{\mathbf{k}} \times \Delta\mathbf{r}_i\right) \\ &= \left(\hat{\mathbf{k}} \times \Delta\mathbf{r}_i\right)\cdot\left(-\Delta\mathbf{r}_i \times \hat{\mathbf{k}}\right) \\ &= -\hat{\mathbf{k}}\cdot\left(\Delta\mathbf{r}_i \times \Delta\mathbf{r}_i \times \hat{\mathbf{k}}\right) \\ &= -\hat{\mathbf{k}}\cdot[\Delta\mathbf{r}_i]^2\,\hat{\mathbf{k}}. \end{aligned}$$

Thus, the moment of inertia around the line L through $\mathbf{R}$ in the direction $\hat{\mathbf{k}}$ is obtained from the calculation,

$$\begin{aligned} I_L &= \sum_{i=1}^{N} m_i |\Delta\mathbf{r}_i^\perp|^2 \\ &= -\sum_{i=1}^{N} m_i \hat{\mathbf{k}}\cdot[\Delta\mathbf{r}_i]^2\,\hat{\mathbf{k}} = \hat{\mathbf{k}}\cdot\left(-\sum_{i=1}^{N} m_i [\Delta\mathbf{r}_i]^2\right)\hat{\mathbf{k}} \\ &= \hat{\mathbf{k}}\cdot\mathbf{I}_\mathbf{R}\hat{\mathbf{k}} = \hat{\mathbf{k}}^\mathsf{T}\mathbf{I}_\mathbf{R}\hat{\mathbf{k}}, \end{aligned}$$

where $\mathbf{I}_\mathbf{R}$ is the moment of inertia matrix of the system relative to the reference point $\mathbf{R}$.

This shows that the inertia matrix can be used to calculate the moment of inertia of a body around any specified rotation axis in the body.

Inertia Tensor

For the same object, different axes of rotation will have different moments of inertia about those axes. In general, the moments of inertia are not equal unless the object is symmetric about all axes.

The moment of inertia tensor is a convenient way to summarize all moments of inertia of an object with one quantity. It may be calculated with respect to any point in space, although for practical purposes the center of mass is most commonly used.

For a rigid object of N point masses m_k, the moment of inertia tensor is given by,

$$\mathbf{I} = \begin{bmatrix} I_{11} & I_{12} & I_{13} \\ I_{21} & I_{22} & I_{23} \\ I_{31} & I_{32} & I_{33} \end{bmatrix}.$$

Its components are defined as,

$$I_{ij} \overset{\text{def}}{=} \sum_{k=1}^{N} m_k (\|\mathbf{r}\|^2 \delta_{ij} - x_i x_j)$$

where

i, j equal 1, 2, or 3 for x, y, and z, respectively,

$\mathbf{r}=(x_1, x_2, x_3)$ is the vector to the mass element dm from the point about which the tensor is calculated, $r=||x||$, and

δ_{ij} is the Kronecker delta.

Note that, by the definition, I is a symmetric tensor.

The diagonal elements, also called the principal moments of inertia, are more succinctly written as,

$$I_{xx} \overset{\text{def}}{=} \sum_{k=1}^{N} m_k (y_k^2 + z_k^2),$$

$$I_{yy} \overset{\text{def}}{=} \sum_{k=1}^{N} m_k (x_k^2 + z_k^2),$$

$$I_{zz} \overset{\text{def}}{=} \sum_{k=1}^{N} m_k (x_k^2 + y_k^2),$$

while the off-diagonal elements, also called the products of inertia, are

$$I_{xy} = I_{yx} \overset{\text{def}}{=} -\sum_{k=1}^{N} m_k x_k y_k,$$

$$I_{xz} = I_{zx} \overset{\text{def}}{=} -\sum_{k=1}^{N} m_k x_k z_k, \text{ and}$$

$$I_{yz} = I_{zy} \overset{\text{def}}{=} -\sum_{k=1}^{N} m_k y_k z_k,$$

Here I_{xx} denotes the moment of inertia around the x-axis when the objects are rotated around the x-axis, I_{xy} denotes the moment of inertia around the y-axis when the objects are rotated around the x-axis, and so on.

These quantities can be generalized to an object with distributed mass, described by a mass density function, in a similar fashion to the scalar moment of inertia. One then has,

$$\mathbf{I} = \iiint_V \rho(x,y,z)\left(\|\mathbf{r}\|^2 \mathbf{E}_3 - \mathbf{r}\otimes\mathbf{r}\right)dx\,dy\,dz,$$

where $\mathbf{r}\otimes\mathbf{r}$ is their outer product, $\mathbf{E}_3$ is the 3 × 3 identity matrix, and V is a region of space completely containing the object.

Alternatively it can also be written in terms of the angular momentum operator $[\mathbf{r}]\mathbf{x} = \mathbf{r}\times\mathbf{x}$:

$$\mathbf{I} = \iiint_V \rho(\mathbf{r})[\mathbf{r}]^T[\mathbf{r}]\mathrm{d}V = -\iiint_Q \rho(\mathbf{r})[\mathbf{r}]^2\,\mathrm{d}V$$

The inertia tensor can be used in the same way as the inertia matrix to compute the scalar moment of inertia about an arbitrary axis in the direction $\mathbf{n}$,

$$I_n = \mathbf{n}\cdot\mathbf{I}\cdot\mathbf{n},$$

where the dot product is taken with the corresponding elements in the component tensors. A product of inertia term such as I_{12} is obtained by the computation,

$$I_{12} = \mathbf{e}_1\cdot\mathbf{I}\cdot\mathbf{e}_2,$$

and can be interpreted as the moment of inertia around the x-axis when the object rotates around the y-axis.

The components of tensors of degree two can be assembled into a matrix. For the inertia tensor this matrix is given by,

$$\mathbf{I} = \begin{bmatrix} I_{11} & I_{12} & I_{13} \\ I_{21} & I_{22} & I_{23} \\ I_{31} & I_{32} & I_{33} \end{bmatrix} = \begin{bmatrix} I_{xx} & I_{xy} & I_{xz} \\ I_{yx} & I_{yy} & I_{yz} \\ I_{zx} & I_{zy} & I_{zz} \end{bmatrix}.$$

It is common in rigid body mechanics to use notation that explicitly identifies the x, y, and z-axes, such as I_{xx} and I_{xy}, for the components of the inertia tensor.

Derivation of the Tensor Components

The distance r of a particle at $\mathbf{x}$ from the axis of rotation passing through the origin in the $\hat{\mathbf{n}}$ direction is $|\mathbf{x} - (\mathbf{x}\cdot\hat{\mathbf{n}})\hat{\mathbf{n}}|$. By using the formula $I = mr^2$ (and some simple vector algebra) it can be seen that the moment of inertia of this particle (about the axis of rotation passing through the origin in the $\hat{\mathbf{n}}$ direction) is $I = m(|\mathbf{x}|^2(\hat{\mathbf{n}}\cdot\hat{\mathbf{n}}) - (\mathbf{x}\cdot\hat{\mathbf{n}})^2)$ This is a quadratic form in $\hat{\mathbf{n}}$ and, after a bit more algebra, this leads to a tensor formula for the moment of inertia,

$$I = m[n_1, n_2, n_3]\begin{bmatrix} y^2+z^2 & -xy & -xz \\ -yx & x^2+z^2 & -yz \\ -zx & -zy & x^2+y^2 \end{bmatrix}\begin{bmatrix} n_1 \\ n_2 \\ n_3 \end{bmatrix}.$$

For multiple particles we need only recall that the moment of inertia is additive in order to see that this formula is correct.

Inertia Matrix in Different Reference Frames

The use of the inertia matrix in Newton's second law assumes its components are computed relative to axes parallel to the inertial frame and not relative to a body-fixed reference frame. This means that as the body moves the components of the inertia matrix change with time. In contrast, the components of the inertia matrix measured in a body-fixed frame are constant.

Body Frame

Let the body frame inertia matrix relative to the center of mass be denoted $\mathbf{I}_C^B$, and define the orientation of the body frame relative to the inertial frame by the rotation matrix $\mathbf{A}$, such that,

$$\mathbf{x} = \mathbf{A}\mathbf{y},$$

where vectors $\mathbf{y}$ in the body fixed coordinate frame have coordinates $\mathbf{x}$ in the inertial frame. Then, the inertia matrix of the body measured in the inertial frame is given by,

$$\mathbf{I}_C = \mathbf{A}\mathbf{I}_C^B\mathbf{A}^\mathsf{T}.$$

Notice that $\mathbf{A}$ changes as the body moves, while $\mathbf{I}_C^B$ remains constant.

Principal Axes

Measured in the body frame the inertia matrix is a constant real symmetric matrix. A real symmetric matrix has the eigendecomposition into the product of a rotation matrix $\mathbf{Q}$ and a diagonal matrix Λ, given by

$$\mathbf{I}_C^B = \mathbf{Q}\Lambda\mathbf{Q}^\mathsf{T},$$

where

$$\Lambda = \begin{bmatrix} I_1 & 0 & 0 \\ 0 & I_2 & 0 \\ 0 & 0 & I_3 \end{bmatrix}.$$

The columns of the rotation matrix $\mathbf{Q}$ define the directions of the principal axes of the body, and the constants I_1, I_2, and I_3 are called the principal moments of inertia. This result was first shown by J. J. Sylvester (1852), and is a form of Sylvester's law of inertia. The principal axis with the highest moment of inertia is sometimes called the figure axis or axis of figure.

When all principal moments of inertia are distinct, the principal axes through center of mass are uniquely specified. If two principal moments are the same, the rigid body is called a symmetrical top and there is no unique choice for the two corresponding principal axes. If all three principal moments are the same, the rigid body is called a spherical top (although it need not be spherical) and any axis can be considered a principal axis, meaning that the moment of inertia is the same about any axis.

The principal axes are often aligned with the object's symmetry axes. If a rigid body has an axis of symmetry of order m, meaning it is symmetrical under rotations of $360°/m$ about the given axis, that axis is a principal axis. When $m > 2$, the rigid body is a symmetrical top. If a rigid body has at least two symmetry axes that are not parallel or perpendicular to each other, it is a spherical top, for example, a cube or any other Platonic solid.

The motion of vehicles is often described in terms of yaw, pitch, and roll which usually correspond approximately to rotations about the three principal axes. If the vehicle has bilateral symmetry then one of the principal axes will correspond exactly to the transverse (pitch) axis.

A practical example of this mathematical phenomenon is the routine automotive task of balancing a tire, which basically means adjusting the distribution of mass of a car wheel such that its principal axis of inertia is aligned with the axle so the wheel does not wobble.

Ellipsoid

The moment of inertia matrix in body-frame coordinates is a quadratic form that defines a surface in the body called Poinsot's ellipsoid. Let Λ be the inertia matrix relative to the center of mass aligned with the principal axes, then the surface,

$$\mathbf{x}^{\mathrm{T}}\Lambda\mathbf{x} = 1,$$

or

$$I_1x^2 + I_2y^2 + I_3z^2 = 1,$$

defines an ellipsoid in the body frame. Write this equation in the form,

$$\left(\frac{x}{1/\sqrt{I_1}}\right)^2 + \left(\frac{y}{1/\sqrt{I_2}}\right)^2 + \left(\frac{z}{1/\sqrt{I_3}}\right)^2 = 1,$$

to see that the semi-principal diameters of this ellipsoid are given by,

$$a = \frac{1}{\sqrt{I_1}}, \quad b = \frac{1}{\sqrt{I_2}}, \quad c = \frac{1}{\sqrt{I_3}}.$$

An ellipsoid with the semi-principal diameters labelled a, b, and c.

Let a point $\mathbf{x}$ on this ellipsoid be defined in terms of its magnitude and direction, $\mathbf{x} = \|\mathbf{x}\|\mathbf{n}$, where $\mathbf{n}$ is a unit vector. Then the relationship presented above, between the inertia matrix and the scalar moment of inertia $I_{\mathbf{n}}$ around an axis in the direction $\mathbf{n}$, yields,

$$\mathbf{x}^{T}\Lambda\mathbf{x} = \|\mathbf{x}\|^{2}\mathbf{n}^{T}\Lambda\mathbf{n} = \|\mathbf{x}\|^{2} I_{\mathbf{n}} = 1.$$

Thus, the magnitude of a point $\mathbf{x}$ in the direction $\mathbf{n}$ on the inertia ellipsoid is,

$$\|\mathbf{x}\| = \frac{1}{\sqrt{I_{\mathbf{n}}}}.$$

Angular Momentum

In physics, angular momentum (rarely, moment of momentum or rotational momentum) is the rotational equivalent of linear momentum. It is an important quantity in physics because it is a conserved quantity—the total angular momentum of a closed system remains constant.

In three dimensions, the angular momentum for a point particle is a pseudovector $\mathbf{r} \times \mathbf{p}$, the cross product of the particle's position vector $\mathbf{r}$ (relative to some origin) and its momentum vector; the latter is $\mathbf{p} = m\mathbf{v}$ in Newtonian mechanics. This definition can be applied to each point in continua like solids or fluids, or physical fields. Unlike momentum, angular momentum does depend on where the origin is chosen, since the particle's position is measured from it.

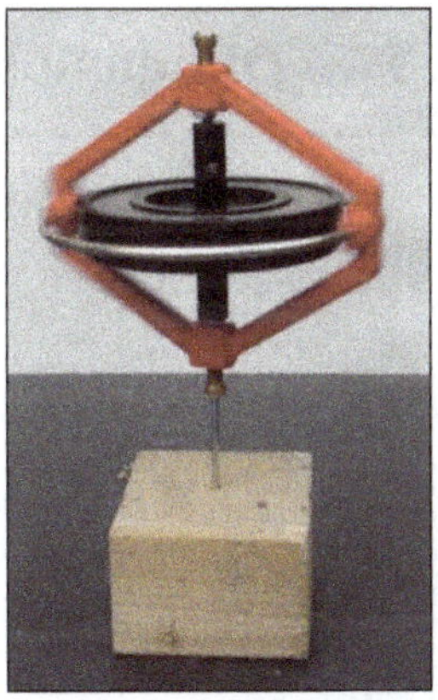

This gyroscope remains upright while spinning due to the conservation of its angular momentum.

Just like for angular velocity, there are two special types of angular momentum: the spin angular momentum and the orbital angular momentum. The spin angular momentum of an object is defined as the angular momentum about its centre of mass coordinate. The orbital angular momentum of an object about a chosen origin is defined as the angular momentum of the centre of mass about the origin. The total angular momentum of an object is the sum of the spin and orbital angular momenta. The orbital angular momentum vector of a particle is always parallel and directly proportional to the orbital angular velocity vector $\boldsymbol{\omega}$ of the particle, where the constant of proportionality depends on both the mass of the particle and its distance from origin. However, the spin angular momentum of the object is proportional but not always parallel to the spin angular velocity $\boldsymbol{\Omega}$, making the constant of proportionality a second-rank tensor rather than a scalar.

Angular momentum is additive; the total angular momentum of any composite system is the (pseudo) vector sum of the angular momenta of its constituent parts. For a continuous rigid body, the total angular momentum is the volume integral of angular momentum density (i.e. angular momentum per unit volume in the limit as volume shrinks to zero) over the entire body.

Torque can be defined as the rate of change of angular momentum, analogous to force. The net *external* torque on any system is always equal to the *total* torque on the system; in other words, the sum of all internal torques of any system is always 0 (this is the rotational analogue of Newton's Third Law). Therefore, for a *closed* system (where there is no net external torque), the *total* torque on the system must be 0, which means that the total angular momentum of the system is constant. The conservation of angular momentum helps explain many observed phenomena, for example the increase in rotational speed of a spinning figure skater as the skater's arms are contracted, the high rotational rates of neutron stars, the Coriolis effect, and the precession of gyroscopes. In general, conservation does limit the possible motion of a system, but does not uniquely determine what the exact motion is.

In quantum mechanics, angular momentum (like other quantities) is expressed as an operator, and its one-dimensional projections have quantized eigenvalues. Angular momentum is subject to the Heisenberg uncertainty principle, implying that at any time, only one projection (also called "component") can be measured with definite precision; the other two then remain uncertain. Because of this, it turns out that the notion of a quantum particle literally "spinning" about an axis does not exist. Nevertheless, elementary particles still possess a spin angular momentum, but this angular momentum does not correspond to spinning motion in the ordinary sense.

Orbital Angular Momentum in Two Dimensions

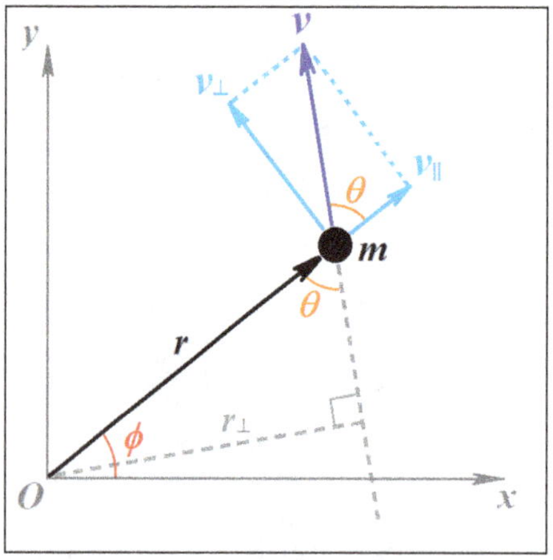

In the figure above, velocity of the particle m with respect to the origin O can be resolved into components parallel to ($v_{\parallel}$) and perpendicular to ($v_{\perp}$) the radius vector r. The angular momentum of m is proportional to the perpendicular component $v_{\perp}$ of the velocity, or equivalently, to the perpendicular distance $r_{\perp}$ from the origin.

Angular momentum is a vector quantity (more precisely, a pseudovector) that represents the product of a body's rotational inertia and rotational velocity (in radians/sec) about a particular axis.

However, if the particle's trajectory lies in a single plane, it is sufficient to discard the vector nature of angular momentum, and treat it as a scalar (more precisely, a pseudoscalar). Angular momentum can be considered a rotational analog of linear momentum. Thus, where linear momentum p is proportional to mass m and linear speed v,

$$p = mv,$$

angular momentum L is proportional to moment of inertia I and angular speed ω measured in radians per second.

$$L = I\omega.$$

Unlike mass, which depends only on amount of matter, moment of inertia is also dependent on the position of the axis of rotation and the shape of the matter. Unlike linear speed, which does not depend upon the choice of origin, angular velocity is always measured with respect to a fixed origin. Therefore, strictly speaking, L should be referred to as the angular momentum relative to that center.

Because $I = r^2 m$ for a single particle and $\omega = \frac{v}{r}$ for circular motion, angular momentum can be expanded, $L = r^2 m \cdot \frac{v}{r}$, and reduced to,

$$L = rmv,$$

the product of the radius of rotation r and the linear momentum of the particle $p = mv$, where v in this case is the equivalent linear (tangential) speed at the radius ($= r\omega$).

This simple analysis can also apply to non-circular motion if only the component of the motion which is perpendicular to the radius vector is considered. In that case,

$$L = rmv_{\perp},$$

where $v_{\perp} = v\sin(\theta)$ is the perpendicular component of the motion. Expanding, $L = rmv\sin(\theta)$, rearranging, $L = r\sin(\theta)mv$, and reducing, angular momentum can also be expressed,

$$L = r_{\perp}mv,$$

where $r_{\perp} = r\sin(\theta)$ is the length of the moment arm, a line dropped perpendicularly from the origin onto the path of the particle. It is this definition, (length of moment arm)×(linear momentum) to which the term moment of momentum refers.

Scalar—Angular Momentum from Lagrangian Mechanics

Another approach is to define angular momentum as the conjugate momentum (also called canonical momentum) of the angular coordinate ϕ expressed in the Lagrangian of the mechanical system. Consider a mechanical system with a mass m constrained to move in a circle of radius a in the absence of any external force field. The kinetic energy of the system is,

$$T = \frac{1}{2}ma^2\omega^2 = \frac{1}{2}ma^2\dot{\phi}^2.$$

And the potential energy is,

$$U = 0.$$

Then the Lagrangian is,

$$\mathcal{L}(\phi, \dot{\phi}) = T - U = \frac{1}{2} ma^2 \dot{\phi}^2.$$

The generalized momentum "canonically conjugate to" the coordinate ϕ is defined by,

$$p_\phi = \frac{\partial \mathcal{L}}{\partial \dot{\phi}} = ma^2 \dot{\phi} = I\omega = L.$$

Orbital Angular Momentum in Three Dimensions

To completely define orbital angular momentum in three dimensions, it is required to know the rate at which the position vector sweeps out angle, the direction perpendicular to the instantaneous plane of angular displacement, and the mass involved, as well as how this mass is distributed in space. By retaining this vector nature of angular momentum, the general nature of the equations is also retained, and can describe any sort of three-dimensional motion about the center of rotation – circular, linear, or otherwise. In vector notation, the orbital angular momentum of a point particle in motion about the origin can be expressed as:

$$\mathbf{L} = I\boldsymbol{\omega},$$

where,

$I = r^2 m$ is the moment of inertia for a point mass,

$\boldsymbol{\omega} = \frac{\mathbf{r} \times \mathbf{v}}{r^2}$ is the orbital angular velocity in radians/sec (units 1/sec) of the particle about the origin,

$\mathbf{r}$ is the position vector of the particle relative to the origin, $r = |\mathbf{r}|$,

$\mathbf{v}$ is the linear velocity of the particle relative to the origin, and

m is the mass of the particle.

This can be expanded, reduced, and by the rules of vector algebra, rearranged:

$$\begin{aligned} \mathbf{L} &= (r^2 m)\left(\frac{\mathbf{r} \times \mathbf{v}}{r^2}\right) \\ &= m(\mathbf{r} \times \mathbf{v}) \\ &= \mathbf{r} \times m\mathbf{v} \\ &= \mathbf{r} \times \mathbf{p}, \end{aligned}$$

which is the cross product of the position vector $\mathbf{r}$ and the linear momentum $\mathbf{p} = m\mathbf{v}$ of the particle. By the definition of the cross product, the $\mathbf{L}$ vector is perpendicular to both $\mathbf{r}$ and $\mathbf{p}$. It is directed perpendicular to the plane of angular displacement, as indicated by the right-hand rule – so that the angular velocity is seen as counter-clockwise from the head of the vector. Conversely, the $\mathbf{L}$ vector defines the plane in which $\mathbf{r}$ and $\mathbf{p}$ lie.

By defining a unit vector $\hat{\mathbf{u}}$ perpendicular to the plane of angular displacement, a scalar angular speed ω results, where

$\omega\hat{\mathbf{u}} = \boldsymbol{\omega}$, and

$\omega = \frac{v_\perp}{r}$, where $v_\perp$ is the perpendicular component of the motion, as above.

The two-dimensional scalar equations can thus be given direction:

$$
\begin{aligned}
\mathbf{L} &= I\boldsymbol{\omega} \\
&= I\omega\hat{\mathbf{u}} \\
&= \left(r^2 m\right)\omega\hat{\mathbf{u}} \\
&= rmv_\perp\hat{\mathbf{u}} \\
&= r_\perp mv\hat{\mathbf{u}},
\end{aligned}
$$

and $\mathbf{L} = rmv\hat{\mathbf{u}}$ for circular motion, where all of the motion is perpendicular to the radius r.

Angular momentum can be described as the rotational analog of linear momentum. Like linear momentum it involves elements of mass and displacement. Unlike linear momentum it also involves elements of position and shape.

Many problems in physics involve matter in motion about some certain point in space, be it in actual rotation about it, or simply moving past it, where it is desired to know what effect the moving matter has on the point—can it exert energy upon it or perform work about it? Energy, the ability to do work, can be stored in matter by setting it in motion—a combination of its inertia and its displacement. Inertia is measured by its mass, and displacement by its velocity. Their product,

$$
aligned(\text{amount of inertia}) \times (\text{amount of displacement}) = \text{amount of }(\text{inertia}\cdot\text{displacement})
$$
$$
\text{mass} \times \text{velocity} = \text{momentum}
$$
$$
m \times v = p
$$

is the matter's momentum. Referring this momentum to a central point introduces a complication: the momentum is not applied to the point directly. For instance, a particle of matter at the outer edge of a wheel is, in effect, at the end of a lever of the same length as the wheel's radius, its momentum turning the lever about the center point. This imaginary lever is known as the *moment arm*. It has the effect of multiplying the momentum's effort in proportion to its length, an effect known as a *moment*. Hence, the particle's momentum referred to a particular point,

$$
(\text{moment arm}) \times (\text{amount of inertia}) \times (\text{amount of displacement}) = \text{moment of }(\text{inertia}\cdot\text{displacement})
$$
$$
\text{length} \times \text{mass} \times \text{velocity} = \text{moment of momentum}
$$
$$
r \times m \times v = L
$$

is the *angular momentum*, sometimes called, as here, the *moment of momentum* of the particle versus that particular center point. The equation $L = rmv$ combines a moment (a mass m turning moment arm r) with a linear (straight-line equivalent) speed v . Linear speed referred to the central point is simply the product of the distance r and the angular speed ω versus the point: $v = r\omega$, another moment. Hence, angular momentum contains a double moment: $L = rmr\omega$. Simplifying slightly, $L = r^2 m\omega$, the quantity $r^2 m$ is the particle's moment of inertia, sometimes called the second moment of mass. It is a measure of rotational inertia.

Because moment of inertia is a crucial part of the spin angular momentum, the latter necessarily includes all of the complications of the former, which is calculated by multiplying elementary bits of the mass by the squares of their distances from the center of rotation. Therefore, the total moment of inertia, and the angular momentum, is a complex function of the configuration of the matter about the center of rotation and the orientation of the rotation for the various bits.

For a rigid body, for instance a wheel or an asteroid, the orientation of rotation is simply the position of the rotation axis versus the matter of the body. It may or may not pass through the center of mass, or it may lie completely outside of the body. For the same body, angular momentum may take a different value for every possible axis about which rotation may take place. It reaches a minimum when the axis passes through the center of mass.

For a collection of objects revolving about a center, for instance all of the bodies of the Solar System, the orientations may be somewhat organized, as is the Solar System, with most of the bodies' axes lying close to the system's axis. Their orientations may also be completely random.

In brief, the more mass and the farther it is from the center of rotation (the longer the moment arm), the greater the moment of inertia, and therefore the greater the angular momentum for a given angular velocity. In many cases the moment of inertia, and hence the angular momentum, can be simplified by,

$$I = k^2 m,$$

where k is the radius of gyration, the distance from the axis at which the entire mass m may be considered as concentrated.

Similarly, for a point mass m the moment of inertia is defined as,

$$I = r^2 m$$

where r is the radius of the point mass from the center of rotation,

and for any collection of particles m_i as the sum,

$$\sum_i I_i = \sum_i r_i^2 m_i$$

Angular momentum's dependence on position and shape is reflected in its units versus linear momentum: kg·m²/s, N·m·s, or J·s for angular momentum versus kg·m/s or N·s for linear momentum. When calculating angular momentum as the product of the moment of inertia times the angular velocity, the angular velocity must be expressed in radians per second, where the radian assumes the dimensionless value of unity. (When performing dimensional analysis, it may be productive to use orientational analysis which treats radians as a base unit, but this is outside the scope of the International system of units). Angular momentum's units can be interpreted as torque·time or as energy·time per angle. An object with angular momentum of L N·m·s can be reduced to zero rotation (all of the rotational energy can be transferred out of it) by an angular impulse of L N·m·s or equivalently, by torque or work of L N·m for one second, or energy of L J for one second.

The plane perpendicular to the axis of angular momentum and passing through the center of mass is sometimes called the *invariable plane*, because the direction of the axis remains fixed if only the interactions of the bodies within the system, free from outside influences, are considered. One such plane is the invariable plane of the Solar System.

Angular Momentum and Torque

Newton's second law of motion can be expressed mathematically,

$$\mathbf{F} = m\mathbf{a},$$

or force = mass × acceleration. The rotational equivalent for point particles may be derived as follows:

$$\mathbf{L} = I\omega$$

which means that the torque (i.e. the time derivative of the angular momentum) is,

$$\tau = \frac{dI}{dt}\omega + I\frac{d\omega}{dt}.$$

Because the moment of inertia is mr^2, it follows that $\frac{dI}{dt} = 2mr\frac{dr}{dt} = 2rp_{\|}$, and $\frac{d\mathbf{L}}{dt} = I\frac{d\omega}{dt} + 2rp_{\|}\omega$, which, reduces to

$$\tau = I\alpha + 2rp_{\|}\omega.$$

This is the rotational analog of Newton's Second Law. Note that the torque is not necessarily proportional or parallel to the angular acceleration (as one might expect). The reason for this is that the moment of inertia of a particle can change with time, something that cannot occur for ordinary mass.

Conservation of Angular Momentum

A rotational analog of Newton's third law of motion might be written, "In a closed system, no torque can be exerted on any matter without the exertion on some other matter of an equal and opposite torque." Hence, angular momentum can be exchanged between objects in a closed system, but total angular momentum before and after an exchange remains constant (is conserved).

Seen another way, a rotational analogue of Newton's first law of motion might be written, "A rigid body continues in a state of uniform rotation unless acted by an external influence." Thus with no external influence to act upon it, the original angular momentum of the system remains constant.

The conservation of angular momentum is used in analyzing *central force motion*. If the net force on some body is directed always toward some point, the *center*, then there is no torque on the body with respect to the center, as all of the force is directed along the radius vector, and none is perpendicular to the radius. Mathematically, torque $\tau = \mathbf{r} \times \mathbf{F} = \mathbf{0}$ because in this case $\mathbf{r}$ and $\mathbf{F}$ are parallel vectors. Therefore, the angular momentum of the body about the center is constant. This is the case with gravitational attraction in the orbits of planets and satellites, where the gravitational force is always directed toward the primary body and orbiting bodies conserve angular momentum by exchanging distance and velocity as they move about the primary. Central force motion is also used in the analysis of the Bohr model of the atom.

For a planet, angular momentum is distributed between the spin of the planet and its revolution in its orbit, and these are often exchanged by various mechanisms. The conservation of angular momentum in the Earth–Moon system results in the transfer of angular momentum from Earth to Moon, due to tidal torque the Moon exerts on the Earth. This in turn results in the slowing down of the rotation rate of Earth, at about 65.7 nanoseconds per day, and in gradual increase of the radius of Moon's orbit, at about 3.82 centimeters per year.

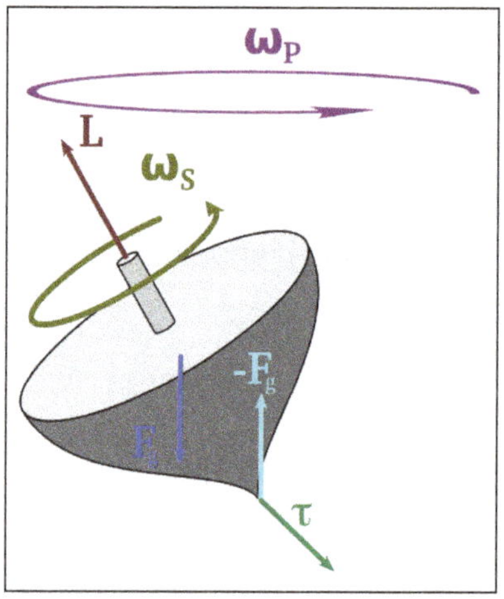

The torque caused by the two opposing forces $\mathbf{F}_g$ and $-\mathbf{F}_g$ causes a change in the angular momentum $\mathbf{L}$ in the direction of that torque (since torque is the time derivative of angular momentum). This causes the top to precess.

The conservation of angular momentum explains the angular acceleration of an ice skater as she brings her arms and legs close to the vertical axis of rotation. By bringing part of the mass of her body closer to the axis, she decreases her body's moment of inertia. Because angular momentum is the product of moment of inertia and angular velocity, if the angular momentum remains constant (is conserved), then the angular velocity (rotational speed) of the skater must increase.

The same phenomenon results in extremely fast spin of compact stars (like white dwarfs, neutron stars and black holes) when they are formed out of much larger and slower rotating stars. Decrease in the size of an object n times results in increase of its angular velocity by the factor of n^2.

Conservation is not always a full explanation for the dynamics of a system but is a key constraint. For example, a spinning top is subject to gravitational torque making it lean over and change the angular momentum about the nutation axis, but neglecting friction at the point of spinning contact, it has a conserved angular momentum about its spinning axis, and another about its precession axis. Also, in any planetary system, the planets, star(s), comets, and asteroids can all move in numerous complicated ways, but only so that the angular momentum of the system is conserved.

Noether's theorem states that every conservation law is associated with a symmetry (invariant) of the underlying physics. The symmetry associated with conservation of angular momentum is rotational invariance. The fact that the physics of a system is unchanged if it is rotated by any angle about an axis implies that angular momentum is conserved.

Angular Momentum in Orbital Mechanics

In astrodynamics and celestial mechanics, a *massless* (or *per unit mass*) angular momentum is defined,

$$\mathbf{h} = \mathbf{r} \times \mathbf{v},$$

called specific angular momentum. Note that $\mathbf{L} = m\mathbf{h}$. Mass is often unimportant in orbital mechanics calculations, because motion is defined by gravity. The primary body of the system is often so much larger than any bodies in motion about it that the smaller bodies have a negligible gravitational effect on it; it is, in effect, stationary. All bodies are apparently attracted by its gravity in the same way, regardless of mass, and therefore all move approximately the same way under the same conditions.

Solid Bodies

For a continuous mass distribution with density function $\rho(\mathbf{r})$, a differential volume element dV with position vector $\mathbf{r}$ within the mass has a mass element $dm = \rho(\mathbf{r})dV$. Therefore, the infinitesimal angular momentum of this element is:

$$d\mathbf{L} = \mathbf{r} \times dm\mathbf{v} = \mathbf{r} \times \rho(\mathbf{r})dV\mathbf{v} = dV\mathbf{r} \times \rho(\mathbf{r})\mathbf{v}$$

and integrating this differential over the volume of the entire mass gives its total angular momentum:

$$\mathbf{L} = \int_V dV\mathbf{r} \times \rho(\mathbf{r})\mathbf{v}$$

In the derivation which follows, integrals similar to this can replace the sums for the case of continuous mass.

Collection of Particles

Center of Mass

For a collection of particles in motion about an arbitrary origin, it is informative to develop the equation of angular momentum by resolving their motion into components about their own center of mass and about the origin. Given,

m_i is the mass of particle i,

$\mathbf{R}_i$ is the position vector of particle i vs the origin,

$\mathbf{V}_i$ is the velocity of particle i vs the origin,

$\mathbf{R}$ is the position vector of the center of mass vs the origin,

$\mathbf{V}$ is the velocity of the center of mass vs the origin,

$\mathbf{r}_i$ is the position vector of particle i vs the center of mass,

$\mathbf{v}_i$ is the velocity of particle i vs the center of mass.

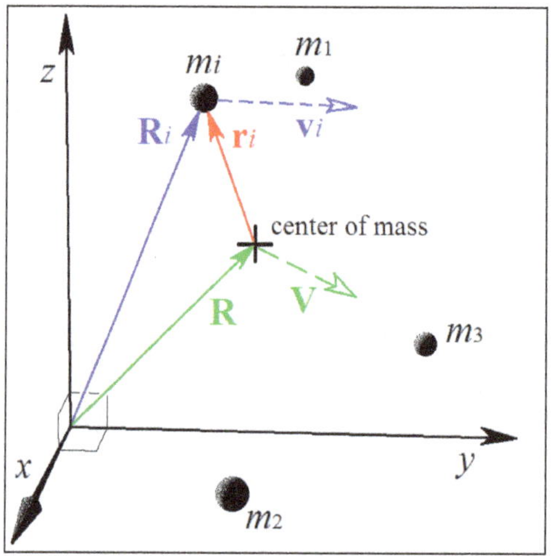

The angular momentum of the particles i is the sum of the cross products $\mathbf{R} \times M\mathbf{V} + \Sigma \mathbf{r}_i \times m_i \mathbf{v}_i$.

The total mass of the particles is simply their sum,

$$M = \sum_i m_i.$$

The position vector of the center of mass is defined by,

$$M\mathbf{R} = \sum_i m_i \mathbf{R}_i.$$

By inspection,

$$\mathbf{R}_i = \mathbf{R} + \mathbf{r}_i \text{ and } \mathbf{V}_i = \mathbf{V} + \mathbf{v}_i.$$

The total angular momentum of the collection of particles is the sum of the angular momentum of each particle,

$$\mathbf{L} = \sum_i (\mathbf{R}_i \times m_i \mathbf{V}_i)$$

Expanding $\mathbf{R}_i$,

$$\mathbf{L} = \sum_i [(\mathbf{R} + \mathbf{r}_i) \times m_i \mathbf{V}_i]$$

$$= \sum_i [\mathbf{R} \times m_i \mathbf{V}_i + \mathbf{r}_i \times m_i \mathbf{V}_i]$$

Expanding $\mathbf{V}_i$,

$$\mathbf{L} = \sum_i \left[\mathbf{R} \times m_i (\mathbf{V} + \mathbf{v}_i) + \mathbf{r}_i \times m_i (\mathbf{V} + \mathbf{v}_i) \right]$$

$$= \sum_i \left[\mathbf{R} \times m_i \mathbf{V} + \mathbf{R} \times m_i \mathbf{v}_i + \mathbf{r}_i \times m_i \mathbf{V} + \mathbf{r}_i \times m_i \mathbf{v}_i \right]$$

$$= \sum_i \mathbf{R} \times m_i \mathbf{V} + \sum_i \mathbf{R} \times m_i \mathbf{v}_i + \sum_i \mathbf{r}_i \times m_i \mathbf{V} + \sum_i \mathbf{r}_i \times m_i \mathbf{v}_i$$

It can be shown that,

$$\sum_i m_i \mathbf{r}_i = \mathbf{0} \text{ and } \sum_i m_i \mathbf{v}_i = \mathbf{0},$$

therefore the second and third terms vanish,

$$\mathbf{L} = \sum_i \mathbf{R} \times m_i \mathbf{V} + \sum_i \mathbf{r}_i \times m_i \mathbf{v}_i.$$

The first term can be rearranged,

$$\sum_i \mathbf{R} \times m_i \mathbf{V} = \mathbf{R} \times \sum_i m_i \mathbf{V} = \mathbf{R} \times M\mathbf{V},$$

and total angular momentum for the collection of particles is finally,

$$\mathbf{L} = \mathbf{R} \times M\mathbf{V} + \sum_i \mathbf{r}_i \times m_i \mathbf{v}_i$$

The first term is the angular momentum of the center of mass relative to the origin. Similar to Single particle, below, it is the angular momentum of one particle of mass M at the center of mass moving with velocity $\mathbf{V}$. The second term is the angular momentum of the particles moving relative to the center of mass, similar to Fixed center of mass, below. The result is general—the motion of the particles is not restricted to rotation or revolution about the origin or center of mass. The particles need not be individual masses, but can be elements of a continuous distribution, such as a solid body.

Rearranging equation $\mathbf{L} = \mathbf{R} \times M\mathbf{V} + \sum_i \mathbf{r}_i \times m_i \mathbf{v}_i$ by vector identities, multiplying both terms by "one", and grouping appropriately,

$$\mathbf{L} = M(\mathbf{R} \times \mathbf{V}) + \sum_i \left[m_i (\mathbf{r}_i \times \mathbf{v}_i) \right],$$

$$= \frac{R^2}{R^2} M (\mathbf{R} \times \mathbf{V}) + \sum_i \left[\frac{r_i^2}{r_i^2} m_i (\mathbf{r}_i \times \mathbf{v}_i) \right],$$

$$= R^2 M \left(\frac{\mathbf{R} \times \mathbf{V}}{R^2} \right) + \sum_i \left[r_i^2 m_i \left(\frac{\mathbf{r}_i \times \mathbf{v}_i}{r_i^2} \right) \right],$$

gives the total angular momentum of the system of particles in terms of moment of inertia I and angular velocity ω,

$$\mathbf{L} = I_R \omega_R + \sum_i I_i \omega_i.$$

Simplifications

Single Particle

In the case of a single particle moving about the arbitrary origin,

$$\mathbf{r}_i = \mathbf{v}_i = \mathbf{0},$$

$$\mathbf{r} = \mathbf{R},$$

$$\mathbf{v} = \mathbf{V},$$

$$m = M,$$

$$\sum_i \mathbf{r}_i \times m_i \mathbf{v}_i = \mathbf{0},$$

$\sum I_i \grave{\mathbf{u}}_i = \mathbf{0}$, and equations $\mathbf{L} = \mathbf{R} \times M\mathbf{V} + \sum_i \mathbf{r}_i \times m_i \mathbf{v}_i$ and $\mathbf{L} = I_R \omega_R + \sum_i I_i \omega_i$ for total angular momentum reduce to,

$$\mathbf{L} = \mathbf{R} \times m\mathbf{V} = I_R \omega_R.$$

Fixed Center of Mass

For the case of the center of mass fixed in space with respect to the origin,

$$\mathbf{V} = \mathbf{0},$$

$$\mathbf{R} \times M\mathbf{V} = \mathbf{0},$$

$I_R \omega_R = \mathbf{0}$, and equations $\mathbf{L} = \mathbf{R} \times M\mathbf{V} + \sum_i \mathbf{r}_i \times m_i \mathbf{v}_i$ and $\mathbf{L} = I_R \omega_R + \sum_i I_i \omega_i$ for total angular momentum reduce to,

$$\mathbf{L} = \sum_i \mathbf{r}_i \times m_i \mathbf{v}_i = \sum_i I_i \omega_i.$$

Torque

Torque, moment, moment of force or "turning effect" is the rotational equivalent of linear force. The concept originated with the studies of Archimedes on the usage of levers. Just as a linear force is a push or a pull, a torque can be thought of as a twist to an object. Another definition of torque is the product of the magnitude of the force and the perpendicular distance of the line of action of force from the axis of rotation. The symbol for torque is typically τ, the lowercase letter *tau*. When being referred to as moment of force, it is commonly denoted by *M*.

In three dimensions, the torque is a pseudovector; for point particles, it is given by the cross product of the position vector (distance vector) and the force vector. The magnitude of torque of a rigid body depends on three quantities: the force applied, the *lever arm vector* connecting the origin to the point of force application, and the angle between the force and lever arm vectors. In symbols:

$$\tau = \mathbf{r} \times \mathbf{F}$$

$$\tau = \|\mathbf{r}\| \|\mathbf{F}\| \sin\theta$$

where,

τ is the torque vector and τ is the magnitude of the torque,

r is the position vector (a vector from the origin of the coordinate system defined to the point where the force is applied),

F is the force vector,

× denotes the cross product, which produces a vector that is perpendicular to both **r** and **F** following the right-hand rule,

θ is the angle between the force vector and the lever arm vector.

The SI unit for torque is N·m.

The term torque was introduced into English scientific literature by James Thomson, the brother of Lord Kelvin, in 1884. However, torque is referred to using different vocabulary depending on geographical location and field of study. This topic refers to the definition used in US physics in its usage of the word torque. In the UK and in US mechanical engineering, torque is referred to as moment of force, usually shortened to moment. In US physics and UK physics terminology these terms are interchangeable, unlike in US mechanical engineering, where the term torque is used for the closely related "resultant moment of a couple".

Torque is defined mathematically as the rate of change of angular momentum of an object. The definition of torque states that one or both of the angular velocity or the moment of inertia of an object are changing. Moment is the general term used for the tendency of one or more applied forces to rotate an object about an axis, but not necessarily to change the angular momentum of the object (the concept which is called torque in physics). For example, a rotational force applied to a shaft causing acceleration, such as a drill bit accelerating from rest, results in a moment called a torque. By contrast, a lateral force on a beam produces a moment (called a bending moment), but since the angular momentum of the beam is not changing, this bending moment is not called a torque. Similarly with any force couple on an object that has no change to its angular momentum, such moment is also not called a torque.

Relation to Angular Momentum

A particle is located at position **r** relative to its axis of rotation, in the figure above. When a force **F** is applied to the particle, only the perpendicular component $\mathbf{F}_\perp$ produces a torque. This torque $\boldsymbol{\tau} = \mathbf{r} \times \mathbf{F}$ has magnitude $\tau = |\mathbf{r}|\,|\mathbf{F}_\perp| = |\mathbf{r}|\,|\mathbf{F}| \sin\theta$.

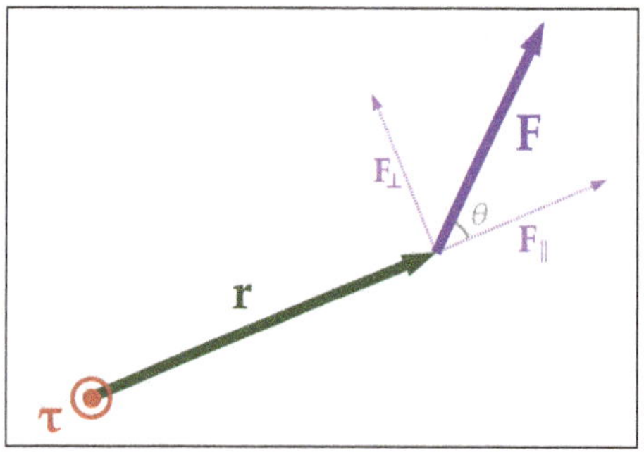

A force applied at a right angle to a lever multiplied by its distance from the lever's fulcrum (the length of the lever arm) is its torque. A force of three newtons applied two metres from the fulcrum, for example, exerts the same torque as a force of one newton applied six metres from the fulcrum. The direction of the torque can be determined by using the right hand grip rule: if the fingers of the right hand are curled from the direction of the lever arm to the direction of the force, then the thumb points in the direction of the torque.

More generally, the torque on a point particle (which has the position **r** in some reference frame) can be defined as the cross product:

$$\tau = \mathbf{r} \times \mathbf{F},$$

where **r** is the particle's position vector relative to the fulcrum, and **F** is the force acting on the particle. The magnitude τ of the torque is given by,

$$\tau = rF\sin\theta,$$

where r is the distance from the axis of rotation to the particle, F is the magnitude of the force applied, and θ is the angle between the position and force vectors. Alternatively,

$$\tau = rF_{\perp},$$

where $F_{\perp}$ is the amount of force directed perpendicularly to the position of the particle. Any force directed parallel to the particle's position vector does not produce a torque.

It follows from the properties of the cross product that the torque vector is perpendicular to both the position and force vectors. Conversely, the torque vector defines the plane in which the position and force vectors lie. The resulting torque vector direction is determined by the right-hand rule.

The net torque on a body determines the rate of change of the body's angular momentum,

$$\tau = \frac{d\mathbf{L}}{dt}$$

where **L** is the angular momentum vector and t is time.

For the motion of a point particle,

$$\mathbf{L} = I\omega,$$

where I is the moment of inertia and $\boldsymbol{\omega}$ is the orbital angular velocity pseudovector. It follows that,

$$\tau_{\text{net}} = \frac{\mathrm{d}\mathbf{L}}{\mathrm{d}t} = \frac{\mathrm{d}(I\omega)}{\mathrm{d}t} = I\frac{\mathrm{d}\omega}{\mathrm{d}t} + \frac{\mathrm{d}I}{\mathrm{d}t}\omega = I\alpha + \frac{\mathrm{d}(mr^2)}{\mathrm{d}t}\omega = I\alpha + 2rp_{||}\omega,$$

where $\boldsymbol{\alpha}$ is the angular acceleration of the particle, and $p_{||}$ is the radial component of its linear momentum. This equation is the rotational analogue of Newton's Second Law for point particles, and is valid for any type of trajectory. Note that although force and acceleration are always parallel and directly proportional, the torque $\boldsymbol{\tau}$ need not be parallel or directly proportional to the angular acceleration $\boldsymbol{\alpha}$. This arises from the fact that although mass is always conserved, the moment of inertia in general is not.

Proof of the Equivalence of Definitions

The definition of angular momentum for a single point particle is:

$$\mathbf{L} = \mathbf{r} \times \mathbf{p}$$

where $\mathbf{p}$ is the particle's linear momentum and $\mathbf{r}$ is the position vector from the origin. The time-derivative of this is:

$$\frac{\mathrm{d}\mathbf{L}}{\mathrm{d}t} = \mathbf{r} \times \frac{\mathrm{d}\mathbf{p}}{\mathrm{d}t} + \frac{\mathrm{d}\mathbf{r}}{\mathrm{d}t} \times \mathbf{p}.$$

This result can easily be proven by splitting the vectors into components and applying the product rule. Now using the definition of force $\mathbf{F} = \frac{\mathrm{d}\mathbf{p}}{\mathrm{d}t}$ (whether or not mass is constant) and the definition of velocity $\frac{\mathrm{d}\mathbf{r}}{\mathrm{d}t} = \mathbf{v}$,

$$\frac{\mathrm{d}\mathbf{L}}{\mathrm{d}t} = \mathbf{r} \times \mathbf{F} + \mathbf{v} \times \mathbf{p}.$$

The cross product of momentum $\mathbf{p}$ with its associated velocity $\mathbf{v}$ is zero because velocity and momentum are parallel, so the second term vanishes.

By definition, torque $\boldsymbol{\tau} = \mathbf{r} \times \mathbf{F}$. Therefore, torque on a particle is *equal* to the first derivative of its angular momentum with respect to time.

If multiple forces are applied, Newton's second law instead reads $\mathbf{F}_{\text{net}} = m\mathbf{a}$, and it follows that,

$$\frac{\mathrm{d}\mathbf{L}}{\mathrm{d}t} = \mathbf{r} \times \mathbf{F}_{\text{net}} = \tau_{\text{net}}.$$

This is a general proof for point particles.

The proof can be generalized to a system of point particles by applying the above proof to each of the point particles and then summing over all the point particles. Similarly, the proof can be generalized to a continuous mass by applying the above proof to each point within the mass, and then integrating over the entire mass.

Units

Torque has the dimension of force times distance, symbolically L^2MT^{-2}. Official SI literature suggests using the unit *newton metre* (N·m), or, *joule per radian* (J/rad). The unit *newton metre* is properly denoted N·m rather than m·N to avoid confusion with mN, millinewtons.

The SI unit for energy or work is the joule. The presence of "joule" in the unit name *joules per radian* for torque is not a coincidence: a torque of 1 N·m applied through a full revolution will require an energy of exactly 2π joules. Mathematically,

$$E = \tau\theta$$

where E is the energy, τ is magnitude of the torque, and θ is the angle moved (in radians). This explains the physical significance of the name *joule per radian*.

Special Cases and other facts

Moment Arm Formula

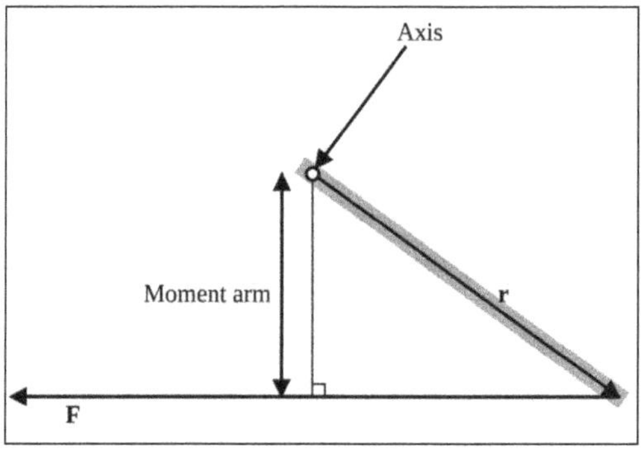

Moment arm diagram

A very useful special case, often given as the definition of torque in fields other than physics, is as follows:

$$\tau = (\text{moment arm})(\text{force}).$$

The construction of the "moment arm" is shown in the figure to the right, along with the vectors **r** and **F** mentioned above. The problem with this definition is that it does not give the direction of the torque but only the magnitude, and hence it is difficult to use in three-dimensional cases. If the force is perpendicular to the displacement vector **r**, the moment arm will be equal to the distance to the centre, and torque will be a maximum for the given force. The equation for the magnitude of a torque, arising from a perpendicular force:

$$\tau = (\text{distance to centre})(\text{force}).$$

For example, if a person places a force of 10 N at the terminal end of a wrench that is 0.5 m long (or a force of 10 N exactly 0.5 m from the twist point of a wrench of any length), the torque will be 5 N·m – assuming that the person moves the wrench by applying force in the plane of movement and perpendicular to the wrench.

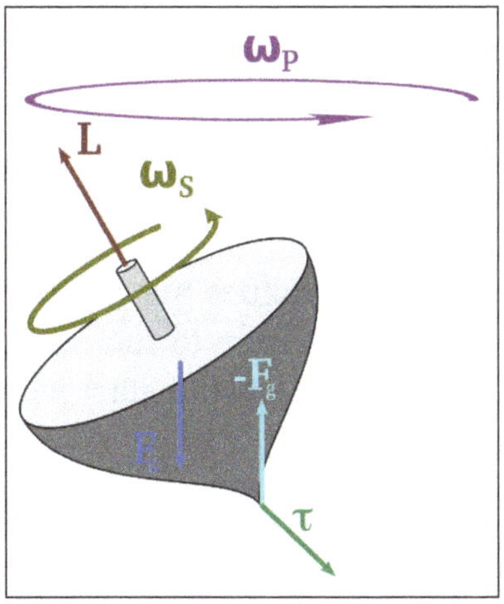

The torque caused by the two opposing forces $\mathbf{F}_g$ and $-\mathbf{F}_g$ causes a change in the angular momentum **L** in the direction of that torque. This causes the top to precess.

Static Equilibrium

For an object to be in static equilibrium, not only must the sum of the forces be zero, but also the sum of the torques (moments) about any point. For a two-dimensional situation with horizontal and vertical forces, the sum of the forces requirement is two equations: $\Sigma H = 0$ and $\Sigma V = 0$, and the torque a third equation: $\Sigma\tau = 0$. That is, to solve statically determinate equilibrium problems in two-dimensions, three equations are used.

Net Force versus Torque

When the net force on the system is zero, the torque measured from any point in space is the same. For example, the torque on a current-carrying loop in a uniform magnetic field is the same regardless of your point of reference. If the net force **F** is not zero, and τ_1 is the torque measured from $\mathbf{r}_1$, then the torque measured from $\mathbf{r}_2$ is ... $\tau_2 = \tau_1 + (\mathbf{r}_1 - \mathbf{r}_2) \times \mathbf{F}$.

Machine Torque

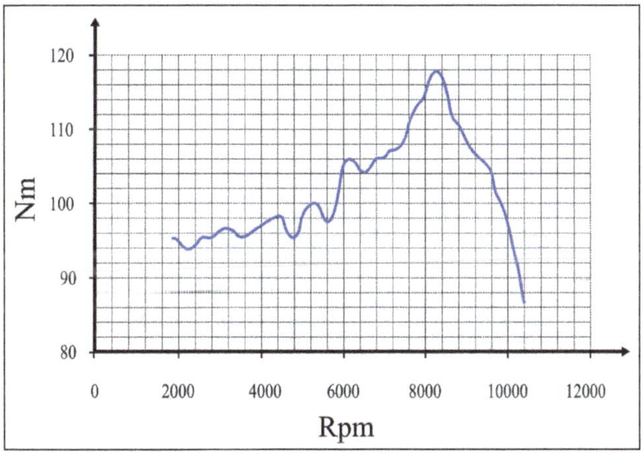

Torque curve of a motorcycle ("BMW K 1200 R 2005"). The horizontal axis shows the speed (in rpm) that the crankshaft is turning, and the vertical axis is the torque (in newton metres) that the engine is capable of providing at that speed.

Torque forms part of the basic specification of an engine: the power output of an engine is expressed as its torque multiplied by its rotational speed of the axis. Internal-combustion engines produce useful torque only over a limited range of rotational speeds (typically from around 1,000–6,000 rpm for a small car). One can measure the varying torque output over that range with a dynamometer, and show it as a torque curve.

Steam engines and electric motors tend to produce maximum torque close to zero rpm, with the torque diminishing as rotational speed rises (due to increasing friction and other constraints). Reciprocating steam-engines and electric motors can start heavy loads from zero rpm without a clutch.

Relationship between Torque, Power and Energy

If a force is allowed to act through a distance, it is doing mechanical work. Similarly, if torque is allowed to act through a rotational distance, it is doing work. Mathematically, for rotation about a fixed axis through the center of mass, the work *W* can be expressed as,

$$W = \int_{\theta_1}^{\theta_2} \tau \, \mathrm{d}\theta,$$

where τ is torque, and θ_1 and θ_2 represent (respectively) the initial and final angular positions of the body.

Proof

The work done by a variable force acting over a finite linear displacement s is given by integrating the force with respect to an elemental linear displacement $\mathrm{d}\vec{s}$,

$$W = \int_{s_1}^{s_2} \vec{F} \cdot \mathrm{d}\vec{s}$$

However, the infinitesimal linear displacement $\mathrm{d}\vec{s}$ is related to a corresponding angular displacement $\mathrm{d}\vec{\theta}$ and the radius vector $\vec{r}$ as,

$$\mathrm{d}\vec{s} = \mathrm{d}\vec{\theta} \times \vec{r}$$

Substitution in the above expression for work gives,

$$W = \int_{s_1}^{s_2} \vec{F} \cdot \mathrm{d}\vec{\theta} \times \vec{r}$$

The expression $\vec{F} \cdot \mathrm{d}\vec{\theta} \times \vec{r}$ is a scalar triple product given by $\left[\vec{F}\mathrm{d}\vec{\theta}\vec{r}\right]$. An alternate expression for the same scalar triple product is,

$$\left[\vec{F}\mathrm{d}\vec{\theta}\vec{r}\right] = \vec{r} \times \vec{F} \cdot \mathrm{d}\vec{\theta}$$

But as per the definition of torque,

$$\vec{\tau} = \vec{r} \times \vec{F}$$

Corresponding substitution in the expression of work gives,

$$W = \int_{s_1}^{s_2} \vec{\tau} \cdot \mathrm{d}\vec{\theta}$$

Since the parameter of integration has been changed from linear displacement to angular displacement, the limits of the integration also change correspondingly, giving:

$$W = \int_{\theta_1}^{\theta_2} \vec{\tau} \cdot \mathrm{d}\vec{\theta}$$

If the torque and the angular displacement are in the same direction, then the scalar product reduces to a product of magnitudes; i.e., $\vec{\tau} \cdot \mathrm{d}\vec{\theta} = |\vec{\tau}||\mathrm{d}\vec{\theta}|\cos 0 = \tau \mathrm{d}\theta$ giving:

$$W = \int_{\theta_1}^{\theta_2} \tau \mathrm{d}\theta$$

It follows from the work-energy theorem that W also represents the change in the rotational kinetic energy E_r of the body, given by:

$$E_r = \tfrac{1}{2} I \omega^2,$$

where I is the moment of inertia of the body and ω is its angular speed.

Power is the work per unit time, given by:

$$P = \tau \cdot \omega,$$

where P is power, τ is torque, ω is the angular velocity, and $\cdot$ represents the scalar product.

Algebraically, the equation may be rearranged to compute torque for a given angular speed and power output. Note that the power injected by the torque depends only on the instantaneous angular speed – not on whether the angular speed increases, decreases, or remains constant while the torque is being applied (this is equivalent to the linear case where the power injected by a force depends only on the instantaneous speed – not on the resulting acceleration, if any).

In practice, this relationship can be observed in bicycles: Bicycles are typically composed of two road wheels, front and rear gears (referred to as sprockets) meshing with a circular chain, and a derailleur mechanism if the bicycle's transmission system allows multiple gear ratios to be used (i.e. multi-speed bicycle), all of which attached to the frame. A cyclist, the person who rides the bicycle, provides the input power by turning pedals, thereby cranking the front sprocket (commonly referred to as chainring). The input power provided by the cyclist is equal to the product of cadence (i.e. the number of pedal revolutions per minute) and the torque on spindle of the bicycle's crankset. The bicycle's drivetrain transmits the input power to the road wheel, which in turn conveys the received power to the road as the output power of the bicycle. Depending on the gear ratio of the bicycle, a $(\text{torque, rpm})_{\text{input}}$ pair is converted to a $(\text{torque, rpm})_{\text{output}}$ pair. By using a larger rear gear, or by switching to a lower gear in multi-speed bicycles, angular speed of the road wheels is decreased while the torque is increased, product of which (i.e. power) does not change.

Consistent units must be used. For metric SI units, power is watts, torque is newton metres and angular speed is radians per second (not rpm and not revolutions per second).

Also, the unit newton metre is dimensionally equivalent to the joule, which is the unit of energy. However, in the case of torque, the unit is assigned to a vector, whereas for energy, it is assigned to a scalar. This means that the dimensional equivalence of the newton metre and the joule may be applied in the former, but not in the latter case. This problem is addressed in orientational analysis which treats radians as a base unit rather than a dimensionless unit.

Conversion to other Units

A conversion factor may be necessary when using different units of power or torque. For example, if rotational speed (revolutions per time) is used in place of angular speed (radians per time), we multiply by a factor of 2π radians per revolution. In the following formulas, P is power, τ is torque, and ν (letter nu) is rotational speed.

$$P = \tau \cdot 2\pi \cdot \nu$$

Showing units:

$$P(\mathrm{W}) = \tau(\mathrm{N{\cdot}m}) \cdot 2\pi(\mathrm{rad/rev}) \cdot \nu(\mathrm{rev/sec})$$

Dividing by 60 seconds per minute gives us the following.

$$P(\mathrm{W}) = \frac{\tau(\mathrm{N{\cdot}m}) \cdot 2\pi(\mathrm{rad/rev}) \cdot \nu(\mathrm{rpm})}{60}$$

where rotational speed is in revolutions per minute (rpm).

Some people (e.g., American automotive engineers) use horsepower (imperial mechanical) for power, foot-pounds (lbf·ft) for torque and rpm for rotational speed. This results in the formula changing to:

$$P(\mathrm{hp}) = \frac{\tau(\mathrm{lbf{\cdot}ft}) \cdot 2\pi(\mathrm{rad/rev}) \cdot \nu(\mathrm{rpm})}{33{,}000}.$$

The constant below (in foot pounds per minute) changes with the definition of the horsepower; for example, using metric horsepower, it becomes approximately 32,550.

Use of other units (e.g., BTU per hour for power) would require a different custom conversion factor.

Derivation

For a rotating object, the *linear distance* covered at the circumference of rotation is the product of the radius with the angle covered. That is: linear distance = radius × angular distance. And by definition, linear distance = linear speed × time = radius × angular speed × time.

By the definition of torque: torque = radius × force. We can rearrange this to determine force = torque ÷ radius. These two values can be substituted into the definition of power:

$$\text{power} = \frac{\text{force} \cdot \text{linear distance}}{\text{time}}$$

$$= \frac{\left(\frac{\text{torque}}{r}\right) \cdot (r \cdot \text{angular speed} \cdot t)}{t}$$

$$= \text{torque} \cdot \text{angular speed}.$$

The radius r and time t have dropped out of the equation. However, angular speed must be in radians, by the assumed direct relationship between linear speed and angular speed at the beginning of the derivation. If the rotational speed is measured in revolutions per unit of time, the linear speed and distance are increased proportionately by 2π in the above derivation to give:

$$\text{power} = \text{torque} \cdot 2\pi \cdot \text{rotational speed}.$$

If torque is in newton metres and rotational speed in revolutions per second, the above equation gives power in newton metres per second or watts. If Imperial units are used, and if torque is in pounds-force feet and rotational speed in revolutions per minute, the above equation gives power in foot pounds-force per minute. The horsepower form of the equation is then derived by applying the conversion factor 33,000 ft·lbf/min per horsepower:

$$\text{power} = \text{torque} \cdot 2\pi \cdot \text{rotational speed} \cdot \frac{\text{ft} \cdot \text{lbf}}{\text{min}} \cdot \frac{\text{horsepower}}{33{,}000 \cdot \frac{\text{ft} \cdot \text{lbf}}{\text{min}}}$$

$$\approx \frac{\text{torque} \cdot \text{RPM}}{5{,}252}$$

because $5252.113122 \approx \frac{33{,}000}{2\pi}$.

Principle of Moments

The Principle of Moments, also known as Varignon's theorem (not to be confused with the geometrical theorem of the same name) states that the sum of torques due to several forces applied to *a single* point is equal to the torque due to the sum (resultant) of the forces. Mathematically, this follows from:

$$(\mathbf{r} \times \mathbf{F}_1) + (\mathbf{r} \times \mathbf{F}_2) + \cdots = \mathbf{r} \times (\mathbf{F}_1 + \mathbf{F}_2 + \cdots).$$

Torque Multiplier

Torque can be multiplied via three methods: by locating the fulcrum such that the length of a lever is increased; by using a longer lever; or by the use of a speed reducing gearset or gear box. Such a mechanism multiplies torque, as rotation rate is reduced.

Rotation Around a Fixed Axis

Rotation around a fixed axis or about a fixed axis of revolution or motion with respect to a fixed axis of rotation is a special case of rotational motion. The fixed axis hypothesis excludes the possibility of an axis changing its orientation, and cannot describe such phenomena as wobbling or precession. According to Euler's rotation theorem, simultaneous rotation along a number of stationary axes at the same time is impossible. If two rotations are forced at the same time, a new axis of rotation will appear.

The kinematics and dynamics of rotation around a fixed axis of a rigid body are mathematically much simpler than those for free rotation of a rigid body; they are entirely analogous to those of linear motion along a single fixed direction, which is not true for *free rotation of a rigid body*. The expressions for the kinetic energy of the object, and for the forces on the parts of the object, are also simpler for rotation around a fixed axis, than for general rotational motion.

Translation and Rotation

A rigid body is an object of finite extent in which all the distances between the component particles are constant. No truly rigid body exists; external forces can deform any solid. For our purposes, then, a rigid body is a solid which requires large forces to deform it appreciably.

A change in the position of a particle in three-dimensional space can be completely specified by three coordinates. A change in the position of a rigid body is more complicated to describe. It can be regarded as a combination of two distinct types of motion: translational motion and circular motion.

Purely translational motion occurs when every particle of the body has the same instantaneous velocity as every other particle; then the path traced out by any particle is exactly parallel to the path traced out by every other particle in the body. Under translational motion, the change in the position of a rigid body is specified completely by three coordinates such as x, y, and z giving the displacement of any point, such as the center of mass, fixed to the rigid body.

Purely rotational motion occurs if every particle in the body moves in a circle about a single line. This line is called the axis of rotation. Then the radius vectors from the axis to all particles undergo the same angular displacement at the same time. The axis of rotation need not go through the body. In general, any rotation can be specified completely by the three angular displacements with respect to the rectangular-coordinate axes x, y, and z. Any change in the position of the rigid body is thus completely described by three translational and three rotational coordinates.

Any displacement of a rigid body may be arrived at by first subjecting the body to a displacement followed by a rotation, or conversely, to a rotation followed by a displacement. We already know that for any collection of particles—whether at rest with respect to one another, as in a rigid body, or in relative motion, like the exploding fragments of a shell, the acceleration of the center of mass is given by,

$$F_{net} = Ma_{cm}$$

where M is the total mass of the system and a_{cm} is the acceleration of the center of mass. There remains the matter of describing the rotation of the body about the center of mass and relating it to the external forces acting on the body. The kinematics and dynamics of *rotational motion around a single axis* resemble the kinematics and dynamics of translational motion; rotational motion around a single axis even has a work-energy theorem analogous to that of particle dynamics.

Kinematics

Angular Displacement

A particle moves in a circle of radius r. Having moved an arc length s, its angular position is θ relative to its original position, where $\theta = \frac{s}{r}$.

In mathematics and physics it is usual to use the natural unit radians rather than degrees or revolutions. Units are converted as follows:

$$1 \text{ revolution } = 360^\circ = 2\pi \text{ radians, and}$$

$$1 \text{ rad} = \frac{180^\circ}{\pi} \approx 57.27^\circ.$$

An angular displacement is a change in angular position:

$$\Delta\theta = \theta_2 - \theta_1,$$

where $\Delta\theta$ is the angular displacement, θ_1 is the initial angular position and θ_2 is the final angular position.

Angular Speed and Angular Velocity

Change in angular displacement per unit time is called angular velocity with direction along the axis of rotation. The symbol for angular velocity is ω and the units are typically rad s^{-1}. Angular speed is the magnitude of angular velocity.

$$\bar{\omega} = \frac{\Delta\theta}{\Delta t} = \frac{\theta_2 - \theta_1}{t_2 - t_1}.$$

The instantaneous angular velocity is given by,

$$\omega(t) = \frac{d\theta}{dt}.$$

Using the formula for angular position and letting $v = \frac{ds}{dt}$, we have also

$$\omega - \frac{d\theta}{dt} = \frac{v}{r},$$

where v is the translational speed of the particle.

Angular velocity and frequency are related by,

$$\omega = 2\pi f.$$

Angular Acceleration

A changing angular velocity indicates the presence of an angular acceleration in rigid body, typically measured in rad s^{-2}. The average angular acceleration $\bar{\alpha}$ over a time interval Δt is given by,

$$\bar{\alpha} = \frac{\Delta\omega}{\Delta t} = \frac{\omega_2 - \omega_1}{t_2 - t_1}.$$

The instantaneous acceleration $\alpha(t)$ is given by,

$$\alpha(t) = \frac{d\omega}{dt} = \frac{d^2\theta}{dt^2}.$$

Thus, the angular acceleration is the rate of change of the angular velocity, just as acceleration is the rate of change of velocity.

The translational acceleration of a point on the object rotating is given by,

$$a = r\alpha,$$

where r is the radius or distance from the axis of rotation. This is also the tangential component of acceleration: it is tangential to the direction of motion of the point. If this component is 0, the motion is uniform circular motion, and the velocity changes in direction only.

The radial acceleration (perpendicular to direction of motion) is given by,

$$a_R = \frac{v^2}{r} = \omega^2 r.$$

It is directed towards the center of the rotational motion, and is often called the centripetal acceleration.

The angular acceleration is caused by the torque, which can have a positive or negative value in accordance with the convention of positive and negative angular frequency. The ratio of torque and angular acceleration (how difficult it is to start, stop, or otherwise change rotation) is given by the moment of inertia: $T = I\alpha$.

Equations of Kinematics

When the angular acceleration is constant, the five quantities angular displacement θ, initial angular velocity ω_i, final angular velocity ω_f, angular acceleration α, and time $t$can be related by four equations of kinematics:

$$\omega_f = \omega_i + \alpha t$$

$$\theta = \omega_i t + \frac{1}{2}\alpha t^2$$

$$\omega_f^2 = \omega_i^2 + 2\alpha\theta$$

$$\theta = \frac{1}{2}\left(\omega_f + \omega_i\right)t$$

Dynamics

Moment of Inertia

The moment of inertia of an object, symbolized by **I**, is a measure of the object's resistance to changes to its rotation. The moment of inertia is measured in kilogram metre^2 (kg m^2). It depends on the object's mass: increasing the mass of an object increases the moment of inertia. It also depends on the distribution of the mass: distributing the mass further from the center of rotation increases the moment of inertia by a greater degree. For a single particle of mass m a distance r from the axis of rotation, the moment of inertia is given by,

$$I = mr^2$$

Torque

Torque τ is the twisting effect of a force **F** applied to a rotating object which is at position **r** from its axis of rotation. Mathematically,

$$\tau = \mathbf{r} \times \mathbf{F},$$

where × denotes the cross product. A net torque acting upon an object will produce an angular acceleration of the object according to,

$$\tau = I\alpha,$$

just as **F** = m**a** in linear dynamics.

The work done by a torque acting on an object equals the magnitude of the torque times the angle through which the torque is applied:

$$W = \tau\theta.$$

The power of a torque is equal to the work done by the torque per unit time, hence:

$$P = \tau\omega.$$

Angular Momentum

The angular momentum **L** is a measure of the difficulty of bringing a rotating object to rest. It is given by,

$$\mathbf{L} = \sum \mathbf{r} \times \mathbf{p} \text{ for all particles in the object.}$$

Angular momentum is the product of moment of inertia and angular velocity:

$$\mathbf{L} = I\omega,$$

just as **p** = m**v** in linear dynamics.

The equivalent of linear momentum in rotational motion is angular momentum. The greater the angular momentum of the spinning object such as a top, the greater its tendency to continue to spin.

The angular momentum of a rotating body is proportional to its mass and to how rapidly it is turning. In addition, the angular momentum depends on how the mass is distributed relative to the axis of rotation: the further away the mass is located from the axis of rotation, the greater the angular momentum. A flat disk such as a record turntable has less angular momentum than a hollow cylinder of the same mass and velocity of rotation.

Like linear momentum, angular momentum is vector quantity, and its conservation implies that the direction of the spin axis tends to remain unchanged. For this reason, the spinning top remains upright whereas a stationary one falls over immediately.

The angular momentum equation can be used to relate the moment of the resultant force on a body about an axis (sometimes called torque), and the rate of rotation about that axis.

Torque and angular momentum are related according to,

$$\tau = \frac{d\mathbf{L}}{dt},$$

just as $\mathbf{F} = d\mathbf{p}/dt$ in linear dynamics. In the absence of an external torque, the angular momentum of a body remains constant. The conservation of angular momentum is notably demonstrated in figure skating: when pulling the arms closer to the body during a spin, the moment of inertia is decreased, and so the angular velocity is increased.

Kinetic Energy

The kinetic energy K_{rot} due to the rotation of the body is given by,

$$K_{rot} = \frac{1}{2} I\omega^2,$$

just as $K_{trans} = \frac{1}{2}mv^2$ in linear dynamics.

Kinetic energy is the energy of motion. The amount of translational kinetic energy found in two variables: the mass of the object (m) and the speed of the object (v) as shown in the equation above. Kinetic energy must always be either zero or a positive value. While velocity can have either a positive or negative value, velocity squared will always be positive.

Vector Expression

The above development is a special case of general rotational motion. In the general case, angular displacement, angular velocity, angular acceleration, and torque are considered to be vectors.

An angular displacement is considered to be a vector, pointing along the axis, of magnitude equal to that of $\Delta\theta$. A right-hand rule is used to find which way it points along the axis; if the fingers of the right hand are curled to point in the way that the object has rotated, then the thumb of the right hand points in the direction of the vector.

The angular velocity vector also points along the axis of rotation in the same way as the angular displacements it causes. If a disk spins counterclockwise as seen from above, its angular velocity vector points upwards. Similarly, the angular acceleration vector points along the axis of rotation in the same direction that the angular velocity would point if the angular acceleration were maintained for a long time.

The torque vector points along the axis around which the torque tends to cause rotation. To maintain rotation around a fixed axis, the total torque vector has to be along the axis, so that it only changes the magnitude and not the direction of the angular velocity vector. In the case of a hinge, only the component of the torque vector along the axis has an effect on the rotation, other forces and torques are compensated by the structure.

Examples and Applications

Constant Angular Speed

The simplest case of rotation around a fixed axis is that of constant angular speed. Then the total torque is zero. For the example of the Earth rotating around its axis, there is very little friction. For a fan, the motor applies a torque to compensate for friction. Similar to the fan, equipment found in the mass production manufacturing industry demonstrate rotation around a fixed axis effectively. For example, a multi-spindle lathe is used to rotate the material on its axis to effectively increase production of cutting, deformation and turning. The angle of rotation is a linear function of time, which modulo 360° is a periodic function.

An example of this is the two-body problem with circular orbits.

Centripetal Force

Internal tensile stress provides the centripetal force that keeps a spinning object together. A rigid body model neglects the accompanying strain. If the body is not rigid this strain will cause it to change shape. This is expressed as the object changing shape due to the “centrifugal force”.

Celestial bodies rotating about each other often have elliptic orbits. The special case of circular orbits is an example of a rotation around a fixed axis: this axis is the line through the center of mass perpendicular to the plane of motion. The centripetal force is provided by gravity. This usually also applies for a spinning celestial body, so it need not be solid to keep together unless the angular speed is too high in relation to its density. (It will, however, tend to become oblate.) For example, a spinning celestial body of water must take at least 3 hours and 18 minutes to rotate, regardless of size, or the water will separate. If the density of the fluid is higher the time can be less. See orbital period.

Rolling

Rolling motion is that common combination of rotational and translational motion that we see everywhere, every day. Think about the different situations of wheels moving on a car along a highway, or wheels on a plane landing on a runway, or wheels on a robotic explorer on another planet. Understanding the forces and torques involved in rolling motion is a crucial factor in many different types of situations.

Rolling Motion without Slipping

In the figure above, (a) The bicycle moves forward, and its tires do not slip. The bottom of the slightly deformed tire is at rest with respect to the road surface for a measurable amount of time. (b) This figure shows that the top of a rolling wheel appears blurred by its motion, but the bottom of the wheel is instantaneously at rest.

People have observed rolling motion without slipping ever since the invention of the wheel. For example, we can look at the interaction of a car's tires and the surface of the road. If the driver depresses the accelerator to the floor, such that the tires spin without the car moving forward, there must be kinetic friction between the wheels and the surface of the road. If the driver depresses the accelerator slowly, causing the car to move forward, then the tires roll without slipping. It is surprising to most people that, in fact, the bottom of the wheel is at rest with respect to the ground, indicating there must be static friction between the tires and the road surface. In Figure, the bicycle is in motion with the rider staying upright. The tires have contact with the road surface, and, even though they are rolling, the bottoms of the tires deform slightly, do not slip, and are at rest with respect to the road surface for a measurable amount of time. There must be static friction between the tire and the road surface for this to be so.

To analyze rolling without slipping, we first derive the linear variables of velocity and acceleration of the center of mass of the wheel in terms of the angular variables that describe the wheel's motion. The situation is shown in figure below.

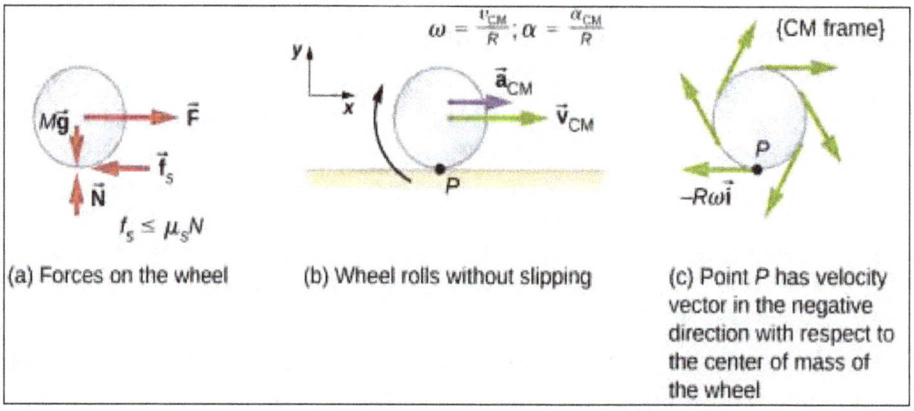

(a) Forces on the wheel

(b) Wheel rolls without slipping

(c) Point *P* has velocity vector in the negative direction with respect to the center of mass of the wheel

In the figure above (a) A wheel is pulled across a horizontal surface by a force $\vec{F}$. The force of static friction $\vec{f}_S, |\vec{f}_S| \le \mu_S$ N is large enough to keep it from slipping. (b) The linear velocity and acceleration vectors of the center of mass and the relevant expressions for ωω and αα. Point P is at rest relative to the surface. (c) Relative to the center of mass (CM) frame, point P has linear velocity $-R\omega\hat{i}$.

From figure above (a), we see the force vectors involved in preventing the wheel from slipping. In (b), point P that touches the surface is at rest relative to the surface. Relative to the center of mass, point P has velocity $-R\omega\hat{i}$, where R is the radius of the wheel and ω is the wheel's angular velocity about its axis. Since the wheel is rolling, the velocity of P with respect to the surface is its velocity with respect to the center of mass plus the velocity of the center of mass with respect to the surface:

$$\vec{v}P = -R\omega\hat{i} + v_{CM}\hat{i}$$

Since the velocity of P relative to the surface is zero, $v_P = 0$, this says that,

$$v_{CM} = R\omega.$$

Thus, the velocity of the wheel's center of mass is its radius times the angular velocity about its axis. We show the correspondence of the linear variable on the left side of the equation with the angular variable on the right side of the equation. This is done below for the linear acceleration.

If we differentiate Equation $\vec{v}P = -R\omega\hat{i} + vCM^{\hat{i}}$ on the left side of the equation, we obtain an expression for the linear acceleration of the center of mass. On the right side of the equation, R is a constant and since $\alpha = \frac{d\omega}{dt}$, we have

$$a_{CM} = R\alpha$$

Furthermore, we can find the distance the wheel travels in terms of angular variables by referring to figure below. As the wheel rolls from point A to point B, its outer surface maps onto the ground by exactly the distance traveled, which is d_{CM}.

We see from figure that the length of the outer surface that maps onto the ground is the arc length Rθ. Equating the two distances, we obtain

$$d_{CM} = R\theta.$$

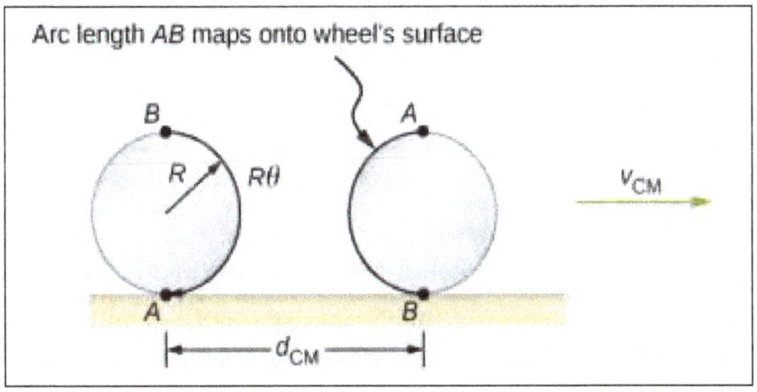

Figure: As the wheel rolls on the surface, the arc length Rθ from A to B maps onto the surface, corresponding to the distance d_{CM} that the center of mass has moved.

It is worthwhile to repeat the equation derived in this example for the acceleration of an object rolling without slipping:

$$a_{CM} = \frac{mg\sin\theta}{m + \left(\frac{Icm}{r^2}\right)}$$

This is a very useful equation for solving problems involving rolling without slipping. Note that the acceleration is less than that for an object sliding down a frictionless plane with no rotation. The acceleration will also be different for two rotating cylinders with different rotational inertias.

Rolling Motion with Slipping

In the case of rolling motion with slipping, we must use the coefficient of kinetic friction, which gives rise to the kinetic friction force since static friction is not present. The situation is shown in figure below. In the case of slipping, $v_{CM} - R\omega \neq 0$, because point P on the wheel is not at rest on the surface, and $v_P \neq 0$. Thus,

$$\omega \neq \frac{v_{CM}}{R}, \alpha \neq \frac{a_{CM}}{R}.$$

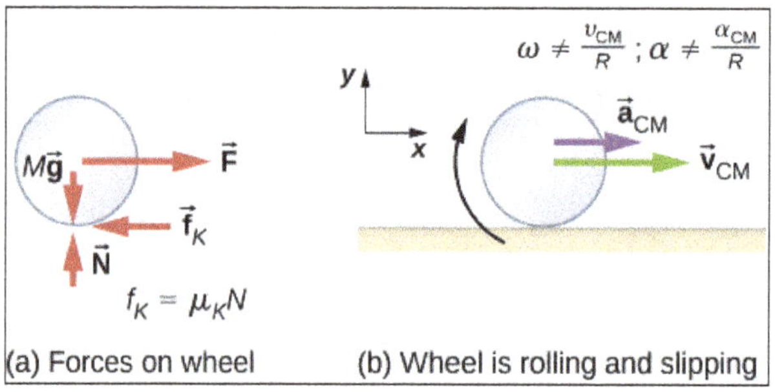

Figure: (a) Kinetic friction arises between the wheel and the surface because the wheel is slipping. (b) The simple relationships between the linear and angular variables are no longer valid.

Conservation of Mechanical Energy in Rolling Motion

Any rolling object carries rotational kinetic energy, as well as translational kinetic energy and potential energy if the system requires. Including the gravitational potential energy, the total mechanical energy of an object rolling is,

$$E_T = \frac{1}{2}mv_{CM}^2 + \frac{1}{2}I_{CM}\omega^2 + mgh.$$

In the absence of any nonconservative forces that would take energy out of the system in the form of heat, the total energy of a rolling object without slipping is conserved and is constant throughout the motion. Examples where energy is not conserved are a rolling object that is slipping, production of heat as a result of kinetic friction, and a rolling object encountering air resistance.

You may ask why a rolling object that is not slipping conserves energy, since the static friction force is nonconservative. The answer can be found by referring back to figure above. Point P in contact with the surface is at rest with respect to the surface. Therefore, its infinitesimal displacement $d\vec{r}$ with respect to the surface is zero, and the incremental work done by the static friction force is zero. We can apply energy conservation to our study of rolling motion to bring out some interesting results.

References

- Ferdinand P. Beer; E. Russell Johnston; Jr., Phillip J. Cornwell (2010). Vector mechanics for engineers: Dynamics (9th ed.). Boston: mcgraw-Hill. ISBN 978-0077295493
- Rotational+motion: encyclopedia2. Thefreedictionary.com, Retrieved 29 March, 2019
- Uicker, John J.; Pennock, Gordon R.; Shigley, Joseph E. (2010). Theory of Machines and Mechanisms (4th ed.). Oxford University Press. ISBN 978-0195371239
- SI brochure Ed. 8, Section 5.1". Bureau International des Poids et Mesures. 2014. Retrieved 2018-07-06
- Mobberley, Martin (2009-03-01). Cataclysmic Cosmic Events and How to Observe Them. Springer Science & Business Media. ISBN 9780387799469

5

Work and Energy

In physics, the work is defined as the product of force and displacement. The capacity of doing work is termed as energy. It includes potential, kinetic, thermal, electrical, chemical and nuclear energy. This chapter closely examines the concepts of work and energy to provide an extensive understanding of the subject.

Work

The baseball pitcher does work on the ball by transferring energy into it.

In physics, mechanical work is the amount of energy transferred by a force. Like energy, it is a scalar quantity, with SI units of joules. Heat conduction is not considered to be a form of work, since there is no macroscopically measurable force, only microscopic forces occurring in atomic collisions. In the 1830s, French mathematician Gaspard-Gustave Coriolis coined the term work for the product of force and distance.

Positive and negative signs of work indicate whether the object exerting the force is transferring energy to some other object, or receiving it. A baseball pitcher, for example, does positive work on the ball, but the catcher does negative work on it. Work can be zero even when there is a force. The centripetal force in uniform circular motion, for example, does zero work because the kinetic energy of the moving object doesn't change. Likewise, when a book sits on a table, the table does no work on the book, because no energy is transferred into or out of the book.

Calculation

When the force is constant and along the same line as the motion, the work can be calculated by multiplying the force by the distance, $W = Fd$ (letting both F and d have positive or negative signs, according to the coordinate system chosen). When the force does not lie along the same line as the motion, this can be generalized to the scalar product of force and displacement vectors.

In the simplest case, that of a body moving in a steady direction, and acted on by a constant force parallel to that direction, the work is given by these formulas:

$$W = FD$$

$$W = (1/2)mv_2^2 - (1/2)mv_1^2 \text{ (derived simply from the above equation)}$$

where

F is the portion of the force acting in the same direction as the motion.

D is the distance traveled by the object. Note that distance is a scalar quantity and so, too, is work.

m is the mass of the object.

v_2 is the final velocity.

v_1 is the initial velocity.

The work is taken to be negative when the force opposes the motion. More generally, the force and distance are taken to be vector quantities, and combined using the dot product:

$$W = \vec{F} \cdot \vec{D} = |\vec{F}||\vec{D}|\cos\phi$$

where ϕ is the angle between the force and the displacement vector. This formula holds true even when the object changes its direction of travel throughout the motion.

In situations in which the force changes over time, and/or the direction of motion changes over time, equation $W = FD$ is not directly applicable. However, under mild restrictions, it is possible to divide the motion into small steps, such that the force and motion are well approximated as being constant for each step, and then to express the overall work as the sum over these steps. This is formalized by the following line integral, which can be taken as a rather general definition of work:

$$W_C := \int_C \vec{F} \cdot d\vec{s}$$

where

C is the path or curve traversed by the object

$\vec{F}$ is the force vector

$\vec{s}$ is the position vector

It must be emphasized that W_C is explicitly a function of the path C. In general the work W_C depends on the total displacement of the path C (The longer the path traveled by the object, the more work will be done).

Using vector notation, equation $W_C := \int_C \vec{F} \cdot d\vec{s}$ readily explains how a nonzero force can do zero work. The simplest case is where the force is always perpendicular to the direction of motion, making the integrand always zero (as is the case in circular motion).

The possibility of a nonzero force doing zero work exemplifies the difference between work and a related quantity, impulse, which is the integral of force over time. Impulse measures change in a body's momentum, a vector quantity sensitive to direction, whereas work considers only the magnitude of the velocity. For instance, as an object in uniform circular motion traverses half of a revolution, its centripetal force does no work, but it transfers a nonzero impulse.

Units

The SI unit of work is the joule (J), which is defined as the work done by a force of one newton acting over a distance of one meter. This definition is based on Sadi Carnot's 1824 definition of work as "weight *lifted* through a height," which is based on the fact that early steam engines were principally used to lift buckets of water, though a gravitational height, out of flooded ore mines. The dimensionally equivalent newton-meter (N•m) is sometimes used instead; however, it is also sometimes reserved for torque to distinguish its units from work or energy.

Non-SI units of work include the erg, the foot-pound, the foot-poundal, and the liter-atmosphere.

Types of Work

Forms of work that are not evidently mechanical in fact represent special cases of this principle. For instance, in the case of "electrical work," an electric field does work on charged particles as they move through a medium.

One mechanism of heat conduction is collisions between fast-moving atoms in a warm body with slow-moving atoms in a cold body. Although colliding atoms do work on each other, the force averages to nearly zero in bulk, so conduction is not considered to be mechanical work.

Work due to Volume Change

Work is done when the volume of a fluid changes. Work in such circumstances is represented by the following equation:

$$W_C = -\int_C P\,\mathrm{d}V$$

where,

W = work done on the system

P = external pressure

V = volume

Like all work functions, PV work is dependent on the path *C*. PV work is often measured in the (non-SI) units of liter-atmospheres, where 1 L•atm = 101.3 J.

Mechanical Energy

The mechanical energy of a body is that part of its total energy which is subject to change by mechanical work. It includes kinetic energy and potential energy. Some notable forms of energy that it does not include are thermal energy (which can be increased by frictional work, but not easily decreased) and rest energy (which is constant as long as the rest mass remains the same).

The Relation between Work and Kinetic Energy

If an external work W acts upon a body, causing its kinetic energy to change from E_{k1} to E_{k2}, then:

$$W = \Delta E_k = E_{k2} - E_{k1}$$

Also, if we substitute the equation for kinetic energy that states $E_k = (1/2)mv^2$, ,we then get:

$$W = \Delta((1/2)mv^2) = (1/2)mv_2^2 - (1/2)mv_1^2$$

Conservation of Mechanical Energy

The principle of conservation of mechanical energy states that, if a system is subject only to conservative forces (such as a gravitational force), its total mechanical energy remains constant.

For instance, if an object with constant mass is in free fall, the total energy of position 1 will equal that of position 2.

$$(K_E + P_E)_1 = (K_E + P_E)_2$$

where

K_E is the kinetic energy, and

P_E is the potential energy.

The external work will usually be done by the friction force between the system on the motion or the internal, non-conservative force in the system, or loss of energy due to heat.

Mathematical Calculation

For moving objects, the quantity of work/time (power) is integrated along the trajectory of the point of application of the force. Thus, at any instant, the rate of the work done by a force (measured in joules/second, or watts) is the scalar product of the force (a vector), and the velocity vector of the point of application. This scalar product of force and velocity is known as instantaneous power. Just as velocities may be integrated over time to obtain a total distance, by the fundamental theorem of calculus, the total work along a path is similarly the time-integral of instantaneous power applied along the trajectory of the point of application.

Work is the result of a force on a point that follows a curve **X**, with a velocity **v**, at each instant. The small amount of work δW that occurs over an instant of time dt is calculated as,

$$\delta W = \mathbf{F} \cdot d\mathbf{s} = \mathbf{F} \cdot \mathbf{v} dt$$

where the $\mathbf{F} \cdot \mathbf{v}$ is the power over the instant dt. The sum of these small amounts of work over the trajectory of the point yields the work,

$$W = \int_{t_1}^{t_2} \mathbf{F} \cdot \mathbf{v} dt = \int_{t_1}^{t_2} \mathbf{F} \cdot \tfrac{d\mathbf{s}}{dt} dt = \int_C \mathbf{F} \cdot d\mathbf{s},$$

where C is the trajectory from $\mathbf{x}(t_1)$ to $\mathbf{x}(t_2)$. This integral is computed along the trajectory of the particle, and is therefore said to be *path dependent*.

If the force is always directed along this line, and the magnitude of the force is F, then this integral simplifies to,

$$W = \int_C F\,ds$$

where s is displacement along the line. If $\mathbf{F}$ is constant, in addition to being directed along the line, then the integral simplifies further to,

$$W = \int_C F\,ds = F\int_C ds = Fs$$

where s is the displacement of the point along the line.

This calculation can be generalized for a constant force that is not directed along the line, followed by the particle. In this case the dot product $\mathbf{F}\cdot d\mathbf{s} = F\cos\theta\, ds$, where θ is the angle between the force vector and the direction of movement, that is

$$W = \int_C \mathbf{F}\cdot d\mathbf{s} = Fs\cos\theta.$$

In the notable case of a force applied to a body always at an angle of 90° from the velocity vector (as when a body moves in a circle under a central force), no work is done at all, since the cosine of 90 degrees is zero. Thus, no work can be performed by gravity on a planet with a circular orbit (this is ideal, as all orbits are slightly elliptical). Also, no work is done on a body moving circularly at a constant speed while constrained by mechanical force, such as moving at constant speed in a frictionless ideal centrifuge.

Work Done by a Variable Force

Calculating the work as "force times straight path segment" would only apply in the most simple of circumstances, as noted above. If force is changing, or if the body is moving along a curved path, possibly rotating and not necessarily rigid, then only the path of the application point of the force is relevant for the work done, and only the component of the force parallel to the application point velocity is doing work (positive work when in the same direction, and negative when in the opposite direction of the velocity). This component of force can be described by the scalar quantity called scalar tangential component ($F\cos\theta$, where θ is the angle between the force and the velocity). And then the most general definition of work can be formulated as follows:

- Work of a force is the line integral of its scalar tangential component along the path of its application point.
- If the force varies (e.g. compressing a spring) we need to use calculus to find the work done. If the force is given by F(x) (a function of x) then the work done by the force along the x-axis from a to b is:

$$W = \int_a^b \mathbf{F}(\mathbf{s})\cdot d\mathbf{s}$$

Torque and Rotation

A force couple results from equal and opposite forces, acting on two different points of a rigid body. The sum (resultant) of these forces may cancel, but their effect on the body is the couple or torque **T**. The work of the torque is calculated as,

$$dW = \mathbf{T}\cdot\vec{\omega}dt,$$

where the $\mathbf{T}\cdot\boldsymbol{\omega}$ is the power over the instant δt. The sum of these small amounts of work over the trajectory of the rigid body yields the work,

$$W = \int_{t_1}^{t_2} \mathbf{T}\cdot\vec{\omega}dt.$$

This integral is computed along the trajectory of the rigid body with an angular velocity ω that varies with time, and is therefore said to be *path dependent.*

If the angular velocity vector maintains a constant direction, then it takes the form,

$$\vec{\omega} = \dot{\phi}\mathbf{S},$$

where φ is the angle of rotation about the constant unit vector **S**. In this case, the work of the torque becomes,

$$W = \int_{t_1}^{t_2} \mathbf{T}\cdot\vec{\omega}dt = \int_{t_1}^{t_2} \mathbf{T}\cdot\mathbf{S}\frac{d\phi}{dt}dt = \int_C \mathbf{T}\cdot\mathbf{S}d\phi,$$

where C is the trajectory from $\varphi(t_1)$ to $\varphi(t_2)$. This integral depends on the rotational trajectory $\varphi(t)$, and is therefore path-dependent.

If the torque **T** is aligned with the angular velocity vector so that,

$$\mathbf{T} = \tau\mathbf{S},$$

and both the torque and angular velocity are constant, then the work takes the form,

$$W = \int_{t_1}^{t_2} \tau\dot{\phi}dt = \tau(\phi_2 - \phi_1).$$

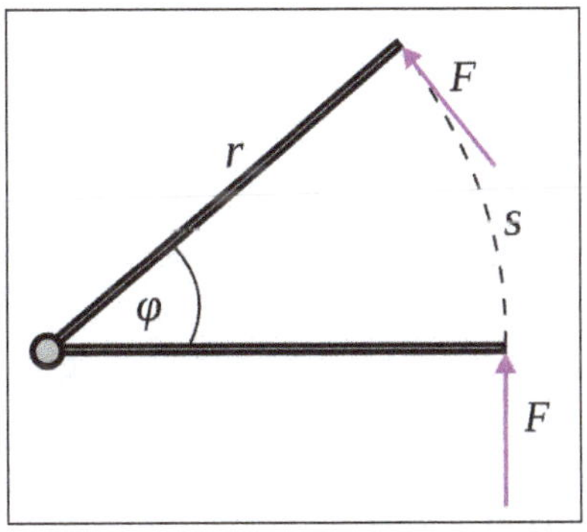

A force of constant magnitude and perpendicular to the lever arm.

This result can be understood more simply by considering the torque as arising from a force of constant magnitude F, being applied perpendicularly to a lever arm at a distance r, as shown in the figure. This force will act through the distance along the circular arc $s = r\varphi$, so the work done is,

$$W = Fs = Fr\phi.$$

Introduce the torque $\tau = Fr$, to obtain,

$$W = Fr\phi = \tau\phi,$$

as presented above.

Notice that only the component of torque in the direction of the angular velocity vector contributes to the work

Work and Potential Energy

The scalar product of a force **F** and the velocity **v** of its point of application defines the power input to a system at an instant of time. Integration of this power over the trajectory of the point of application, $C = \mathbf{x}(t)$, defines the work input to the system by the force.

Path Dependence

Therefore, the work done by a force **F** on an object that travels along a curve C is given by the line integral:

$$W = \int_C \mathbf{F}\cdot d\mathbf{x} = \int_{t_1}^{t_2} \mathbf{F}\cdot\mathbf{v}dt,$$

where $dx(t)$ defines the trajectory C and **v** is the velocity along this trajectory. In general this integral requires the path along which the velocity is defined, so the evaluation of work is said to be path dependent.

The time derivative of the integral for work yields the instantaneous power,

$$\frac{dW}{dt} = P(t) = \mathbf{F}\cdot\mathbf{v}.$$

Path Independence

If the work for an applied force is independent of the path, then the work done by the force, by the gradient theorem, defines a potential function which is evaluated at the start and end of the trajectory of the point of application. This means that there is a potential function $U(\mathbf{x})$, that can be evaluated at the two points $\mathbf{x}(t_1)$ and $\mathbf{x}(t_2)$ to obtain the work over any trajectory between these two points. It is tradition to define this function with a negative sign so that positive work is a reduction in the potential, that is

$$W = \int_C \mathbf{F}\cdot d\mathbf{x} = \int_{\mathbf{x}(t_1)}^{\mathbf{x}(t_2)} \mathbf{F}\cdot d\mathbf{x} = U(\mathbf{x}(t_1)) - U(\mathbf{x}(t_2)).$$

The function $U(\mathbf{x})$ is called the potential energy associated with the applied force. The force derived from such a potential function is said to be conservative. Examples of forces that have potential energies are gravity and spring forces.

In this case, the gradient of work yields,

$$\nabla W = -\nabla U = -\left(\frac{\partial U}{\partial x}, \frac{\partial U}{\partial y}, \frac{\partial U}{\partial z}\right) = \mathbf{F},$$

and the force **F** is said to be "derivable from a potential."

Because the potential U defines a force **F** at every point **x** in space, the set of forces is called a force field. The power applied to a body by a force field is obtained from the gradient of the work, or potential, in the direction of the velocity **V** of the body, that is

$$P(t) = -\nabla U \cdot \mathbf{v} = \mathbf{F} \cdot \mathbf{v}.$$

Work by Gravity

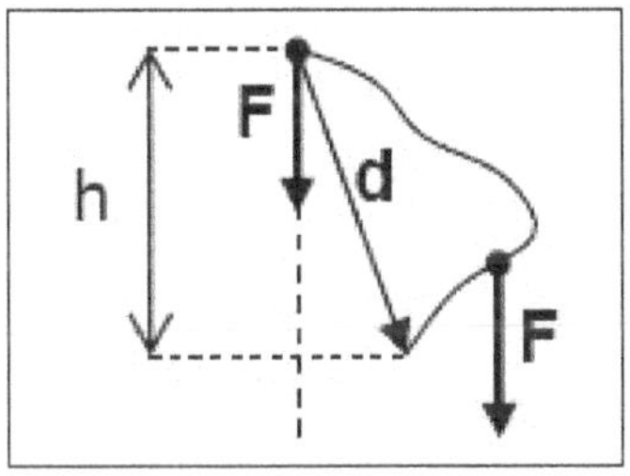

Gravity $F = mg$ does work $W = mgh$ along any descending path

In the absence of other forces, gravity results in a constant downward acceleration of every freely moving object. Near Earth's surface the acceleration due to gravity is $g = 9.8\ \mathrm{m{\cdot}s^{-2}}$ and the gravitational force on an object of mass m is $\mathbf{F}_g = mg$. It is convenient to imagine this gravitational force concentrated at the center of mass of the object.

If an object is displaced upwards or downwards a vertical distance $y_2 - y_1$, the work W done on the object by its weight mg is:

$$W = F_g (y_2 - y_1) = F_g \Delta y = -mg\Delta y$$

where F_g is weight (pounds in imperial units, and newtons in SI units), and Δy is the change in height y. Notice that the work done by gravity depends only on the vertical movement of the object. The presence of friction does not affect the work done on the object by its weight.

Work by Gravity in Space

The force of gravity exerted by a mass M on another mass m is given by,

$$\mathbf{F} = -\frac{GMm}{r^3}\mathbf{r},$$

where **r** is the position vector from M to m.

Let the mass m move at the velocity $\mathbf{v}$ then the work of gravity on this mass as it moves from position $\mathbf{r}(t_1)$ to $\mathbf{r}(t_2)$ is given by,

$$W = -\int_{\mathbf{r}(t_1)}^{\mathbf{r}(t_2)} \frac{GMm}{r^3}\mathbf{r}\cdot d\mathbf{r} = -\int_{t_1}^{t_2} \frac{GMm}{r^3}\mathbf{r}\cdot\mathbf{v}dt.$$

Notice that the position and velocity of the mass m are given by,

$$\mathbf{r} = r\mathbf{e}_r, \qquad \mathbf{v} = \dot{r}\mathbf{e}_r + r\dot{\theta}\mathbf{e}_t,$$

where $\mathbf{e}_r$ and $\mathbf{e}_t$ are the radial and tangential unit vectors directed relative to the vector from M to m. Use this to simplify the formula for work of gravity to,

$$W = -\int_{t_1}^{t_2} \frac{GmM}{r^3}(r\mathbf{e}_r)\cdot(\dot{r}\mathbf{e}_r + r\dot{\theta}\mathbf{e}_t)dt = -\int_{t_1}^{t_2} \frac{GmM}{r^3}r\dot{r}dt = \frac{GMm}{r(t_2)} - \frac{GMm}{r(t_1)}.$$

This calculation uses the fact that,

$$\frac{d}{dt}r^{-1} = -r^{-2}\dot{r} = -\frac{\dot{r}}{r^2}.$$

The function,

$$U = -\frac{GMm}{r},$$

is the gravitational potential function, also known as gravitational potential energy. The negative sign follows the convention that work is gained from a loss of potential energy.

Work by a Spring

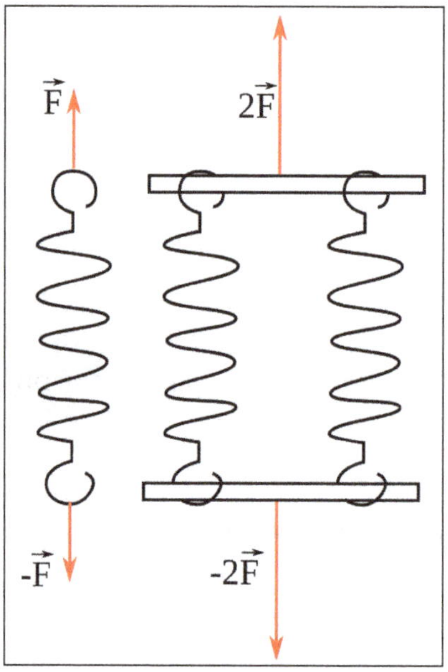

Forces in springs assembled in parallel.

Consider a spring that exerts a horizontal force $\mathbf{F} = (-kx, 0, 0)$ that is proportional to its deflection in the x direction independent of how a body moves. The work of this spring on a body moving along the space with the curve $\mathbf{X}(t) = (x(t), y(t), z(t))$, is calculated using its velocity, $\mathbf{v} = (v_x, v_y, v_z)$, to obtain,

$$W = \int_0^t \mathbf{F}\cdot\mathbf{v}dt = -\int_0^t kxv_x dt = -\frac{1}{2}kx^2.$$

For convenience, consider contact with the spring occurs at $t = 0$, then the integral of the product of the distance x and the x-velocity, xv_x, is $(1/2)x^2$. The velocity is not a factor here. The work is the product of the distance times the spring force, which is also dependent on distance; hence the x^2 result.

Work by a Gas

$$W = \int_a^b PdV$$

Where P is pressure, V is volume, and a and b are initial and final volumes.

Work–energy Principle

The principle of work and kinetic energy (also known as the work–energy principle) states that the work done by all forces acting on a particle (the work of the resultant force) equals the change in the kinetic energy of the particle. That is, the work W done by the resultant force on a particle equals the change in the particle's kinetic energy E_k,

$$W = \Delta E_k = \tfrac{1}{2}mv_2^2 - \tfrac{1}{2}mv_1^2,$$

where v_1 and v_2 are the speeds of the particle before and after the work is done, and m is its mass.

The derivation of the work–energy principle begins with Newton's second law of motion and the resultant force on a particle. Computation of the scalar product of the forces with the velocity of the particle evaluates the instantaneous power added to the system.

Constraints define the direction of movement of the particle by ensuring there is no component of velocity in the direction of the constraint force. This also means the constraint forces do not add to the instantaneous power. The time integral of this scalar equation yields work from the instantaneous power, and kinetic energy from the scalar product of velocity and acceleration. The fact the work–energy principle eliminates the constraint forces underlies Lagrangian mechanics.

This topic focuses on the work–energy principle as it applies to particle dynamics. In more general systems work can change the potential energy of a mechanical device, the thermal energy in a thermal system, or the electrical energy in an electrical device. Work transfers energy from one place to another or one form to another.

Derivation for a Particle Moving along a Straight Line

In the case the resultant force **F** is constant in both magnitude and direction, and parallel to the velocity of the particle, the particle is moving with constant acceleration a along a straight line. The relation between the net force and the acceleration is given by the equation $F = ma$ (Newton's second law), and the particle displacement s can be expressed by the equation,

$$s = \frac{v_2^2 - v_1^2}{2a}$$

which follows from $v_2^2 = v_1^2 + 2as$.

The work of the net force is calculated as the product of its magnitude and the particle displacement. Substituting the above equations, one obtains:

$$W = Fs = mas = ma\left(\frac{v_2^2 - v_1^2}{2a}\right) = \frac{mv_2^2}{2} - \frac{mv_1^2}{2} = \Delta E_k$$

Other derivation:

$$W = Fs = mas = m\left(\frac{v_2^2 - v_1^2}{2s}\right)s$$

$$W = m\left(\frac{0_2^2 - v_1^2}{2s}\right)s$$

$$W = -\frac{1}{2}mv^2$$

Vertical displacement derivation:

$$W = F \times S = mg \times h$$

In the general case of rectilinear motion, when the net force **F** is not constant in magnitude, but is constant in direction, and parallel to the velocity of the particle, the work must be integrated along the path of the particle:

$$W = \int_{t_1}^{t_2} \mathbf{F}\cdot\mathbf{v}dt = \int_{t_1}^{t_2} F\,vdt = \int_{t_1}^{t_2} mavdt = m\int_{t_1}^{t_2} v\frac{dv}{dt}dt = m\int_{v_1}^{v_2} vdv = \tfrac{1}{2}m(v_2^2 - v_1^2).$$

General Derivation of the Work–energy Theorem for a Particle

For any net force acting on a particle moving along any curvilinear path, it can be demonstrated that its work equals the change in the kinetic energy of the particle by a simple derivation analogous to the equation above. Some authors call this result work–energy principle, but it is more widely known as the work–energy theorem:

$$W = \int_{t_1}^{t_2} \mathbf{F}\cdot\mathbf{v}dt = m\int_{t_1}^{t_2} \mathbf{a}\cdot\mathbf{v}dt = \frac{m}{2}\int_{t_1}^{t_2} \frac{dv^2}{dt}dt = \frac{m}{2}\int_{v_1^2}^{v_2^2} dv^2 = \frac{mv_2^2}{2} - \frac{mv_1^2}{2} = \Delta E_k$$

The identity $\mathbf{a}\cdot\mathbf{v} = \frac{1}{2}\frac{dv^2}{dt}$ requires some algebra. From the identity $v^2 = \mathbf{v}\cdot\mathbf{v}$ and definition $\mathbf{a} = \frac{d\mathbf{v}}{dt}$ it follows,

$$\frac{dv^2}{dt} = \frac{d(\mathbf{v}\cdot\mathbf{v})}{dt} = \frac{d\mathbf{v}}{dt}\cdot\mathbf{v} + \mathbf{v}\cdot\frac{d\mathbf{v}}{dt} = 2\frac{d\mathbf{v}}{dt}\cdot\mathbf{v} = 2\mathbf{a}\cdot\mathbf{v}.$$

The remaining part of the above derivation is just simple calculus, same as in the preceding rectilinear case.

Derivation for a Particle in Constrained Movement

In particle dynamics, a formula equating work applied to a system to its change in kinetic energy is obtained as a first integral of Newton's second law of motion. It is useful to notice that the resultant force used in Newton's laws can be separated into forces that are applied to the particle and forces imposed by constraints on the movement of the particle. Remarkably, the work of a constraint force is zero, therefore only the work of the applied forces need be considered in the work–energy principle.

To see this, consider a particle P that follows the trajectory $\mathbf{X}(t)$ with a force $\mathbf{F}$ acting on it. Isolate the particle from its environment to expose constraint forces $\mathbf{R}$, then Newton's Law takes the form,

$$\mathbf{F} + \mathbf{R} = m\ddot{\mathbf{X}},$$

where m is the mass of the particle.

Vector Formulation

Note that n dots above a vector indicates its nth time derivative. The scalar product of each side of Newton's law with the velocity vector yields,

$$\mathbf{F}\cdot\dot{\mathbf{X}} = m\ddot{\mathbf{X}}\cdot\dot{\mathbf{X}},$$

because the constraint forces are perpendicular to the particle velocity. Integrate this equation along its trajectory from the point $\mathbf{X}(t_1)$ to the point $\mathbf{X}(t_2)$ to obtain,

$$\int_{t_1}^{t_2} \mathbf{F}\cdot\dot{\mathbf{X}}\,dt = m\int_{t_1}^{t_2} \ddot{\mathbf{X}}\cdot\dot{\mathbf{X}}\,dt.$$

The left side of this equation is the work of the applied force as it acts on the particle along the trajectory from time t_1 to time t_2. This can also be written as,

$$W = \int_{t_1}^{t_2} \mathbf{F}\cdot\dot{\mathbf{X}}\,dt = \int_{\mathbf{X}(t_1)}^{\mathbf{X}(t_2)} \mathbf{F}\cdot d\mathbf{X}.$$

This integral is computed along the trajectory $\mathbf{X}(t)$ of the particle and is therefore path dependent.

The right side of the first integral of Newton's equations can be simplified using the following identity,

$$\frac{1}{2}\frac{d}{dt}(\dot{\mathbf{X}}\cdot\dot{\mathbf{X}}) = \ddot{\mathbf{X}}\cdot\dot{\mathbf{X}},$$

Now it is integrated explicitly to obtain the change in kinetic energy,

$$\Delta K = m\int_{t_1}^{t_2} \ddot{\mathbf{X}}\cdot\dot{\mathbf{X}}\,dt = \frac{m}{2}\int_{t_1}^{t_2}\frac{d}{dt}(\dot{\mathbf{X}}\cdot\dot{\mathbf{X}})dt = \frac{m}{2}\dot{\mathbf{X}}\cdot\dot{\mathbf{X}}(t_2) - \frac{m}{2}\dot{\mathbf{X}}\cdot\dot{\mathbf{X}}(t_1) = \frac{1}{2}m\Delta\mathbf{v}^2,$$

where the kinetic energy of the particle is defined by the scalar quantity,

$$K = \frac{m}{2}\dot{\mathbf{X}}\cdot\dot{\mathbf{X}} = \frac{1}{2}m\mathbf{v}^2.$$

Tangential and Normal Components

It is useful to resolve the velocity and acceleration vectors into tangential and normal components along the trajectory $\mathbf{X}(t)$, such that

$$\dot{\mathbf{X}} = v\mathbf{T} \quad \text{and} \quad \ddot{\mathbf{X}} = \dot{v}\mathbf{T} + v^2\kappa\mathbf{N},$$

where

$$v = |\dot{\mathbf{X}}| = \sqrt{\dot{\mathbf{X}}\cdot\dot{\mathbf{X}}}.$$

Then, the scalar product of velocity with acceleration in Newton's second law takes the form,

$$\Delta K = m\int_{t_1}^{t_2} \dot{v}v\,dt = \frac{m}{2}\int_{t_1}^{t_2}\frac{d}{dt}v^2\,dt = \frac{m}{2}v^2(t_2) - \frac{m}{2}v^2(t_1),$$

where the kinetic energy of the particle is defined by the scalar quantity,

$$K = \frac{m}{2}v^2 = \frac{m}{2}\dot{\mathbf{X}}\cdot\dot{\mathbf{X}}.$$

The result is the work–energy principle for particle dynamics,

$$W = \Delta K.$$

This derivation can be generalized to arbitrary rigid body systems.

Moving in a Straight Line (Skid to a Stop)

Consider the case of a vehicle moving along a straight horizontal trajectory under the action of a driving force and gravity that sum to $\mathbf{F}$. The constraint forces between the vehicle and the road define $\mathbf{R}$, and we have,

$$\mathbf{F} + \mathbf{R} = m\ddot{\mathbf{X}}.$$

For convenience let the trajectory be along the X-axis, so $\mathbf{X} = (d, 0)$ and the velocity is $\mathbf{V} = (v, 0)$, then $\mathbf{R}\cdot\mathbf{V} = 0$, and $\mathbf{F}\cdot\mathbf{V} = F_x v$, where F_x is the component of $\mathbf{F}$ along the X-axis, so

$$F_x v = m\dot{v}v.$$

Integration of both sides yields,

$$\int_{t_1}^{t_2} F_x v dt = \frac{m}{2}v^2(t_2) - \frac{m}{2}v^2(t_1).$$

If F_x is constant along the trajectory, then the integral of velocity is distance, so

$$F_x(d(t_2) - d(t_1)) = \frac{m}{2}v^2(t_2) - \frac{m}{2}v^2(t_1).$$

As an example consider a car skidding to a stop, where k is the coefficient of friction and W is the weight of the car. Then the force along the trajectory is $F_x = -kW$. The velocity v of the car can be determined from the length s of the skid using the work–energy principle,

$$kWs = \frac{W}{2g}v^2, \quad \text{or} \quad v = \sqrt{2ksg}.$$

Notice that this formula uses the fact that the mass of the vehicle is $m = W/g$.

Coasting Down a Mountain Road (Gravity Racing)

Lotus type 119B gravity racer.

Gravity racing championship.

Consider the case of a vehicle that starts at rest and coasts down a mountain road, the work-energy principle helps compute the minimum distance that the vehicle travels to reach a velocity V, of say 60 mph (88 fps). Rolling resistance and air drag will slow the vehicle down so the actual distance will be greater than if these forces are neglected.

Let the trajectory of the vehicle following the road be $\mathbf{X}(t)$ which is a curve in three-dimensional space. The force acting on the vehicle that pushes it down the road is the constant force of gravity $\mathbf{F} = (0, 0, W)$, while the force of the road on the vehicle is the constraint force $\mathbf{R}$. Newton's second law yields,

$$\mathbf{F} + \mathbf{R} = m\ddot{\mathbf{X}}.$$

The scalar product of this equation with the velocity, $\mathbf{V} = (v_x, v_y, v_z)$, yields:

$$Wv_z = m\dot{V}V,$$

where V is the magnitude of **V**. The constraint forces between the vehicle and the road cancel from this equation because $\mathbf{R} \cdot \mathbf{V} = 0$, which means they do no work. Integrate both sides to obtain,

$$\int_{t_1}^{t_2} W v_z dt = \frac{m}{2}V^2(t_2) - \frac{m}{2}V^2(t_1).$$

The weight force W is constant along the trajectory and the integral of the vertical velocity is the vertical distance, therefore,

$$W\Delta z = \frac{m}{2}V^2.$$

Recall that $V(t_1)=0$. Notice that this result does not depend on the shape of the road followed by the vehicle.

In order to determine the distance along the road assume the downgrade is 6%, which is a steep road. This means the altitude decreases 6 feet for every 100 feet traveled—for angles this small the sin and tan functions are approximately equal. Therefore, the distance s in feet down a 6% grade to reach the velocity V is at least,

$$s = \frac{\Delta z}{0.06} = 8.3\frac{V^2}{g}, \quad \text{or} \quad s = 8.3\frac{88^2}{32.2} \approx 2000\text{ft}.$$

This formula uses the fact that the weight of the vehicle is $W = mg$.

Work of Forces Acting on a Rigid Body

The work of forces acting at various points on a single rigid body can be calculated from the work of a resultant force and torque. To see this, let the forces $\mathbf{F}_1, \mathbf{F}_2 \ldots \mathbf{F}_n$ act on the points $\mathbf{X}_1, \mathbf{X}_2 \ldots \mathbf{X}_n$ in a rigid body.

The trajectories of $\mathbf{X}_i$, $i = 1, \ldots, n$ are defined by the movement of the rigid body. This movement is given by the set of rotations $[A(t)]$ and the trajectory $\mathbf{d}(t)$ of a reference point in the body. Let the coordinates $\mathbf{x}_i$ $i = 1, \ldots, n$ define these points in the moving rigid body's reference frame M, so that the trajectories traced in the fixed frame F are given by,

$$\mathbf{X}_i(t) = [A(t)]\mathbf{x}_i + \mathbf{d}(t) \quad i = 1, \ldots, n.$$

The velocity of the points $\mathbf{X}_i$ along their trajectories are,

$$\mathbf{V}_i = \vec{\omega} \times (\mathbf{X}_i - \mathbf{d}) + \dot{\mathbf{d}},$$

where $\boldsymbol{\omega}$ is the angular velocity vector obtained from the skew symmetric matrix,

$$[\Omega] = \dot{A}A^{\mathrm{T}},$$

known as the angular velocity matrix.

The small amount of work by the forces over the small displacements $\delta\mathbf{r}_i$ can be determined by approximating the displacement by $\delta\mathbf{r} = \mathbf{v}\delta t$ so,

$$\delta W = \mathbf{F}_1 \cdot \mathbf{V}_1 \delta t + \mathbf{F}_2 \cdot \mathbf{V}_2 \delta t + \ldots + \mathbf{F}_n \cdot \mathbf{V}_n \delta t$$

or

$$\delta W = \sum_{i=1}^{n} \mathbf{F}_i \cdot (\vec{\omega} \times (\mathbf{X}_i - \mathbf{d}) + \dot{\mathbf{d}}) \delta t.$$

This formula can be rewritten to obtain,

$$\delta W = \left(\sum_{i=1}^{n} \mathbf{F}_i \right) \cdot \dot{\mathbf{d}} \delta t + \left(\sum_{i=1}^{n} (\mathbf{X}_i - \mathbf{d}) \times \mathbf{F}_i \right) \cdot \vec{\omega} \delta t = \left(\mathbf{F} \cdot \dot{\mathbf{d}} + \mathbf{T} \cdot \vec{\omega} \right) \delta t,$$

where $\mathbf{F}$ and $\mathbf{T}$ are the resultant force and torque applied at the reference point $\mathbf{d}$ of the moving frame M in the rigid body.

Energy

In physics, energy is the capacity for doing work. It may exist in potential, kinetic, thermal, electrical, chemical, nuclear, or other various forms. There are, moreover, heat and work—i.e., energy in the process of transfer from one body to another. After it has been transferred, energy is always designated according to its nature. Hence, heat transferred may become thermal energy, while work done may manifest itself in the form of mechanical energy.

All forms of energy are associated with motion. For example, any given body has kinetic energy if it is in motion. A tensioned device such as a bow or spring, though at rest, has the potential for creating motion; it contains potential energy because of its configuration. Similarly, nuclear energy is potential energy because it results from the configuration of subatomic particles in the nucleus of an atom.

Energy can be neither created nor destroyed but only changed from one form to another. This principle is known as the conservation of energy or the first law of thermodynamics. For example, when a box slides down a hill, the potential energy that the box has from being located high up on the slope is converted to kinetic energy, energy of motion. As the box slows to a stop through friction, the kinetic energy from the box's motion is converted to thermal energy that heats the box and the slope.

Energy can be converted from one form to another in various other ways. Usable mechanical or electrical energy is, for instance, produced by many kinds of devices, including fuel-burning heat engines, generators, batteries, fuel cells, and magnetohydrodynamic systems.

In the International System of Units (SI), energy is measured in joules. One joule is equal to the work done by a one-newton force acting over a one-metre distance.

Kinetic Energy

The cars of a roller coaster reach their maximum kinetic energy when at the bottom of the path. When they start rising, the kinetic energy begins to be converted to gravitational potential energy. The sum of kinetic and potential energy in the system remains constant, ignoring losses to friction.

In physics, the kinetic energy (KE) of an object is the energy that it possesses due to its motion. It is defined as the work needed to accelerate a body of a given mass from rest to its stated velocity. Having gained this energy during its acceleration, the body maintains this kinetic energy unless its speed changes. The same amount of work is done by the body when decelerating from its current speed to a state of rest.

In classical mechanics, the kinetic energy of a non-rotating object of mass m traveling at a speed v is $\frac{1}{2}mv^2$. In relativistic mechanics, this is a good approximation only when v is much less than the speed of light.

The standard unit of kinetic energy is the joule, while the imperial unit of kinetic energy is the foot-pound.

Energy occurs in many forms, including chemical energy, thermal energy, electromagnetic radiation, gravitational energy, electric energy, elastic energy, nuclear energy, and rest energy. These can be categorized in two main classes: potential energy and kinetic energy. Kinetic energy is the movement energy of an object. Kinetic energy can be transferred between objects and transformed into other kinds of energy.

Kinetic energy may be best understood by examples that demonstrate how it is transformed to and from other forms of energy. For example, a cyclist uses chemical energy provided by food to accelerate a bicycle to a chosen speed. On a level surface, this speed can be maintained without further work, except to overcome air resistance and friction. The chemical energy has been converted into kinetic energy, the energy of motion, but the process is not completely efficient and produces heat within the cyclist.

The kinetic energy in the moving cyclist and the bicycle can be converted to other forms. For example, the cyclist could encounter a hill just high enough to coast up, so that the bicycle comes to a complete halt at the top. The kinetic energy has now largely been converted to gravitational potential energy that can be released by freewheeling down the other side of the hill. Since the bicycle lost some of its energy to friction, it never regains all of its speed without additional pedaling. The energy is not destroyed; it has only been converted to another form by friction. Alternatively, the cyclist could connect a dynamo to one of the wheels and generate some electrical energy on the descent. The bicycle would be traveling slower at the bottom of the hill than without the generator because some of the energy has been diverted into electrical energy. Another possibility would be for the cyclist to apply the brakes, in which case the kinetic energy would be dissipated through friction as heat.

Like any physical quantity that is a function of velocity, the kinetic energy of an object depends on the relationship between the object and the observer's frame of reference. Thus, the kinetic energy of an object is not invariant.

Spacecraft use chemical energy to launch and gain considerable kinetic energy to reach orbital velocity. In an entirely circular orbit, this kinetic energy remains constant because there is almost no friction in near-earth space. However, it becomes apparent at re-entry when some of the kinetic energy is converted to heat. If the orbit is elliptical or hyperbolic, then throughout the orbit kinetic and potential energy are exchanged; kinetic energy is greatest and potential energy lowest at closest approach to the earth or other massive body, while potential energy is greatest and kinetic energy the lowest at maximum distance. Without loss or gain, however, the sum of the kinetic and potential energy remains constant.

Kinetic energy can be passed from one object to another. In the game of billiards, the player imposes kinetic energy on the cue ball by striking it with the cue stick. If the cue ball collides with another ball, it slows down dramatically, and the ball it hit accelerates its speed as the kinetic energy is passed on to it. Collisions in billiards are effectively elastic collisions, in which kinetic energy is preserved. In inelastic collisions, kinetic energy is dissipated in various forms of energy, such as heat, sound, binding energy (breaking bound structures).

Flywheels have been developed as a method of energy storage. This illustrates that kinetic energy is also stored in rotational motion.

Several mathematical descriptions of kinetic energy exist that describe it in the appropriate physical situation. For objects and processes in common human experience, the formula $\frac{1}{2}mv^2$ given by Newtonian (classical) mechanics is suitable. However, if the speed of the object is comparable to the speed of light, relativistic effects become significant and the relativistic formula is used. If the object is on the atomic or sub-atomic scale, quantum mechanical effects are significant, and a quantum mechanical model must be employed.

Newtonian Kinetic Energy

Kinetic Energy of Rigid Bodies

In classical mechanics, the kinetic energy of a *point object* (an object so small that its mass can be assumed to exist at one point), or a non-rotating rigid body depends on the mass of the body as well as its speed. The kinetic energy is equal to 1/2 the product of the mass and the square of the speed. In formula form:

$$E_{\mathrm{k}} = \tfrac{1}{2}mv^2$$

where m is the mass and v is the speed (or the velocity) of the body. In SI units, mass is measured in kilograms, speed in metres per second, and the resulting kinetic energy is in joules.

For example, one would calculate the kinetic energy of an 80 kg mass (about 180 lbs) traveling at 18 metres per second (about 40 mph, or 65 km/h) as

$$E_{\mathrm{k}} = \frac{1}{2} \cdot 80\mathrm{kg} \cdot (18\mathrm{m/s})^2 = 12{,}960\mathrm{J} = 12.96\mathrm{kJ}$$

When a person throws a ball, the person does work on it to give it speed as it leaves the hand. The moving ball can then hit something and push it, doing work on what it hits. The kinetic energy of a moving object is equal to the work required to bring it from rest to that speed, or the work the object can do while being brought to rest: net force × displacement = kinetic energy, i.e.,

$$Fs = \tfrac{1}{2}mv^2$$

Since the kinetic energy increases with the square of the speed, an object doubling its speed has four times as much kinetic energy. For example, a car traveling twice as fast as another requires four times as much distance to stop, assuming a constant braking force. As a consequence of this quadrupling, it takes four times the work to double the speed.

The kinetic energy of an object is related to its momentum by the equation:

$$E_{\mathrm{k}} = \frac{p^2}{2m}$$

where,

p is momentum

m is mass of the body

For the *translational kinetic energy,* that is the kinetic energy associated with rectilinear motion, of a rigid body with constant mass m, whose center of mass is moving in a straight line with speed v, as seen above is equal to,

$$E_{\mathrm{t}} = \tfrac{1}{2}mv^2$$

where,

m is the mass of the body

v is the speed of the center of mass of the body.

The kinetic energy of any entity depends on the reference frame in which it is measured. However the total energy of an isolated system, i.e. one in which energy can neither enter nor leave, does not change over time in the reference frame in which it is measured. Thus, the chemical energy converted to kinetic energy by a rocket engine is divided differently between the rocket ship and its exhaust stream depending upon the chosen reference frame. This is called the Oberth effect. But the total energy of the system, including kinetic energy, fuel chemical energy, heat, etc., is conserved over time, regardless of the choice of reference frame. Different observers moving with different reference frames would however disagree on the value of this conserved energy.

The kinetic energy of such systems depends on the choice of reference frame: the reference frame that gives the minimum value of that energy is the center of momentum frame, i.e. the reference frame in which the total momentum of the system is zero. This minimum kinetic energy contributes to the invariant mass of the system as a whole.

Derivation

The work done in accelerating a particle with mass m during the infinitesimal time interval dt is given by the dot product of *force* $\mathbf{F}$ and the infinitesimal *displacement* $\mathbf{dx}$,

$$\mathbf{F}\cdot d\mathbf{x} = \mathbf{F}\cdot\mathbf{v}dt = \frac{d\mathbf{p}}{dt}\cdot\mathbf{v}dt = \mathbf{v}\cdot d\mathbf{p} = \mathbf{v}\cdot d(m\mathbf{v}),$$

where we have assumed the relationship $\mathbf{p} = m\,\mathbf{v}$ and the validity of Newton's Second Law. Applying the product rule we see that:

$$d(\mathbf{v}\cdot\mathbf{v}) = (d\mathbf{v})\cdot\mathbf{v} + \mathbf{v}\cdot(d\mathbf{v}) = 2(\mathbf{v}\cdot d\mathbf{v}).$$

Therefore, (assuming constant mass so that dm=0), we have,

$$\mathbf{v}\cdot d(m\mathbf{v}) = \frac{m}{2}d(\mathbf{v}\cdot\mathbf{v}) = \frac{m}{2}dv^2 = d\left(\frac{mv^2}{2}\right).$$

Since this is a total differential (that is, it only depends on the final state, not how the particle got there), we can integrate it and call the result kinetic energy. Assuming the object was at rest at time 0, we integrate from time 0 to time t because the work done by the force to bring the object from rest to velocity v is equal to the work necessary to do the reverse:

$$E_\text{k} = \int_0^t \mathbf{F}\cdot d\mathbf{x} = \int_0^t \mathbf{v}\cdot d(m\mathbf{v}) = \int_0^v d\left(\frac{mv^2}{2}\right) = \frac{mv^2}{2}.$$

This equation states that the kinetic energy (E_k) is equal to the integral of the dot product of the velocity ($\mathbf{v}$) of a body and the infinitesimal change of the body's momentum ($\mathbf{p}$). It is assumed that the body starts with no kinetic energy when it is at rest (motionless).

Rotating Bodies

If a rigid body Q is rotating about any line through the center of mass then it has rotational kinetic energy (E_r) which is simply the sum of the kinetic energies of its moving parts, and is thus given by:

$$E_\text{r} = \int_Q \frac{v^2 dm}{2} = \int_Q \frac{(r\omega)^2 dm}{2} = \frac{\omega^2}{2}\int_Q r^2 dm = \frac{\omega^2}{2}I = \frac{1}{2}I\omega^2$$

where

ω is the body's angular velocity

r is the distance of any mass dm from that line

I is the body's moment of inertia, equal to $\int_Q r^2 dm$.

(In this equation the moment of inertia must be taken about an axis through the center of mass and the rotation measured by ω must be around that axis; more general equations exist for systems where the object is subject to wobble due to its eccentric shape).

Kinetic Energy of Systems

A system of bodies may have internal kinetic energy due to the relative motion of the bodies in the system. For example, in the Solar System the planets and planetoids are orbiting the Sun. In a tank of gas, the molecules are moving in all directions. The kinetic energy of the system is the sum of the kinetic energies of the bodies it contains.

A macroscopic body that is stationary (i.e. a reference frame has been chosen to correspond to the body's center of momentum) may have various kinds of internal energy at the molecular or atomic level, which may be regarded as kinetic energy, due to molecular translation, rotation, and vibration, electron translation and spin, and nuclear spin. These all contribute to the body's mass, as provided by the special theory of relativity. When discussing movements of a macroscopic body, the kinetic energy referred to is usually that of the macroscopic movement only. However all internal energies of all types contribute to body's mass, inertia, and total energy.

Fluid Dynamics

In fluid dynamics, the kinetic energy per unit volume at each point in an incompressible fluid flow field is called the dynamic pressure at that point.

$$E_{\mathrm{k}} = \tfrac{1}{2} m v^2$$

Dividing by V, the unit of volume:

$$\frac{E_{\mathrm{k}}}{V} = \tfrac{1}{2} \tfrac{m}{V} v^2$$

$$q = \tfrac{1}{2} \rho v^2$$

where q is the dynamic pressure, and ρ is the density of the incompressible fluid.

Frame of Reference

The speed, and thus the kinetic energy of a single object is frame-dependent (relative): it can take any non-negative value, by choosing a suitable inertial frame of reference. For example, a bullet passing an observer has kinetic energy in the reference frame of this observer. The same bullet is stationary to an observer moving with the same velocity as the bullet, and so has zero kinetic energy. By contrast, the total kinetic energy of a system of objects cannot be reduced to zero by a suitable choice of the inertial reference frame, unless all the objects have the same velocity. In any other case, the total kinetic energy has a non-zero minimum, as no inertial reference frame can be chosen in which all the objects are stationary. This minimum kinetic energy contributes to the system's invariant mass, which is independent of the reference frame.

The total kinetic energy of a system depends on the inertial frame of reference: it is the sum of the total kinetic energy in a center of momentum frame and the kinetic energy the total mass would have if it were concentrated in the center of mass.

This may be simply shown: let **V** be the relative velocity of the center of mass frame i in the frame k. Since

$$v^2 = (v_i + V)^2 = (\mathbf{v}_i + \mathbf{V})\cdot(\mathbf{v}_i + \mathbf{V}) = \mathbf{v}_i\cdot\mathbf{v}_i + 2\mathbf{v}_i\cdot\mathbf{V} + \mathbf{V}\cdot\mathbf{V} = v_i^2 + 2\mathbf{v}_i\cdot\mathbf{V} + V^2,$$

Then,

$$E_\text{k} = \int \frac{v^2}{2} dm = \int \frac{v_i^2}{2} dm + \mathbf{V}\cdot\int \mathbf{v}_i dm + \frac{V^2}{2}\int dm.$$

However, let $\int \frac{v_i^2}{2} dm = E_i$ the kinetic energy in the center of mass frame, $\int \mathbf{v}_i dm$ would be simply the total momentum that is by definition zero in the center of mass frame, and let the total mass: $\int dm = M$. Substituting, we get:

$$E_\text{k} = E_i + \frac{MV^2}{2}.$$

Thus the kinetic energy of a system is lowest to center of momentum reference frames, i.e., frames of reference in which the center of mass is stationary (either the center of mass frame or any other center of momentum frame). In any different frame of reference, there is additional kinetic energy corresponding to the total mass moving at the speed of the center of mass. The kinetic energy of the system in the center of momentum frame is a quantity that is invariant (all observers see it to be the same).

Rotation in Systems

It sometimes is convenient to split the total kinetic energy of a body into the sum of the body's center-of-mass translational kinetic energy and the energy of rotation around the center of mass (rotational energy):

$$E_\text{k} = E_t + E_\text{r}$$

where,

E_k is the total kinetic energy

E_t is the translational kinetic energy

E_r is the rotational energy or angular kinetic energy in the rest frame.

Thus the kinetic energy of a tennis ball in flight is the kinetic energy due to its rotation, plus the kinetic energy due to its translation.

Relativistic Kinetic Energy of Rigid Bodies

If a body's speed is a significant fraction of the speed of light, it is necessary to use relativistic mechanics to calculate its kinetic energy. In special relativity theory, the expression for linear momentum is modified.

With m being an object's rest mass, **v** and v its velocity and speed, and c the speed of light in vacuum, we use the expression for linear momentum $\mathbf{p} = m\gamma\mathbf{v}$,

where,

$$\gamma = 1/\sqrt{1 - v^2/c^2}.$$

Integrating by parts yields

$$E_{\text{k}} = \int \mathbf{v} \cdot d\mathbf{p} = \int \mathbf{v} \cdot d(m\gamma\mathbf{v}) = m\gamma\mathbf{v}\cdot\mathbf{v} - \int m\gamma\mathbf{v}\cdot d\mathbf{v} = m\gamma v^2 - \frac{m}{2}\int \gamma d(v^2)$$

Since $\gamma = (1 - v^2/c^2)^{-1/2}$,

$$\begin{aligned} E_{\text{k}} &= m\gamma v^2 - \frac{-mc^2}{2}\int \gamma d(1 - v^2/c^2) \\ &= m\gamma v^2 + mc^2(1 - v^2/c^2)^{1/2} - E_0 \end{aligned}$$

E_0 is a constant of integration for the indefinite integral. Simplifying the expression we obtain,

$$\begin{aligned} E_{\text{k}} &= m\gamma(v^2 + c^2(1 - v^2/c^2)) - E_0 \\ &= m\gamma(v^2 + c^2 - v^2) - E_0 \\ &= m\gamma c^2 - E_0 \end{aligned}$$

E_0 is found by observing that when $\mathbf{v} = 0, \gamma = 1$ and $E_{\text{k}} = 0$, giving,

$$E_0 = mc^2$$

resulting in the formula

$$E_{\text{k}} = m\gamma c^2 - mc^2 = \frac{mc^2}{\sqrt{1 - v^2/c^2}} - mc^2$$

This formula shows that the work expended accelerating an object from rest approaches infinity as the velocity approaches the speed of light. Thus it is impossible to accelerate an object across this boundary.

The mathematical by-product of this calculation is the mass-energy equivalence formula—the body at rest must have energy content,

$$E_{\text{rest}} = E_0 = mc^2$$

At a low speed ($v << c$), the relativistic kinetic energy is approximated well by the classical kinetic energy. This is done by binomial approximation or by taking the first two terms of the Taylor expansion for the reciprocal square root:

$$E_k \approx mc^2\left(1+\frac{1}{2}v^2/c^2\right)-mc^2=\frac{1}{2}mv^2$$

So, the total energy E_k can be partitioned into the rest mass energy plus the Newtonian kinetic energy at low speeds.

When objects move at a speed much slower than light (e.g. in everyday phenomena on Earth), the first two terms of the series predominate. The next term in the Taylor series approximation,

$$E_k \approx mc^2\left(1+\frac{1}{2}v^2/c^2+\frac{3}{8}v^4/c^4\right)-mc^2=\frac{1}{2}mv^2+\frac{3}{8}mv^4/c^2$$

is small for low speeds. For example, for a speed of 10 km/s (22,000 mph) the correction to the Newtonian kinetic energy is 0.0417 J/kg (on a Newtonian kinetic energy of 50 MJ/kg) and for a speed of 100 km/s it is 417 J/kg (on a Newtonian kinetic energy of 5 GJ/kg).

The relativistic relation between kinetic energy and momentum is given by,

$$E=\sqrt{p^2c^2+m^2c^4}-mc^2$$

This can also be expanded as a Taylor series, the first term of which is the simple expression from Newtonian mechanics:

$$E_k \approx \frac{p^2}{2m}-\frac{p^4}{8m^3c^2}.$$

This suggests that the formulae for energy and momentum are not special and axiomatic, but concepts emerging from the equivalence of mass and energy and the principles of relativity.

Potential Energy

In physics, potential energy is the energy held by an object because of its position relative to other objects, stresses within itself, its electric charge, or other factors.

Common types of potential energy include the gravitational potential energy of an object that depends on its mass and its distance from the center of mass of another object, the elastic potential energy of an extended spring, and the electric potential energy of an electric charge in an electric field. The unit for energy in the International System of Units (SI) is the joule, which has the symbol J.

Potential energy is associated with forces that act on a body in a way that the total work done by these forces on the body depends only on the initial and final positions of the body in space. These forces, that are called *conservative forces*, can be represented at every point in space by vectors expressed as gradients of a certain scalar function called *potential*.

Since the work of potential forces acting on a body that moves from a start to an end position is determined only by these two positions, and does not depend on the trajectory of the body, there is a function known as potential that can be evaluated at the two positions to determine this work.

There are various types of potential energy, each associated with a particular type of force. For example, the work of an elastic force is called elastic potential energy; work of the gravitational force is called gravitational potential energy; work of the Coulomb force is called electric potential energy; work of the strong nuclear force or weak nuclear force acting on the baryon charge is called nuclear potential energy; work of intermolecular forces is called intermolecular potential energy. Chemical potential energy, such as the energy stored in fossil fuels, is the work of the Coulomb force during rearrangement of mutual positions of electrons and nuclei in atoms and molecules. Thermal energy usually has two components: the kinetic energy of random motions of particles and the potential energy of their mutual positions.

Forces derivable from a potential are also called conservative forces. The work done by a conservative force is,

$$W = -\Delta U$$

where ΔU is the change in the potential energy associated with the force. The negative sign provides the convention that work done against a force field increases potential energy, while work done by the force field decreases potential energy. Common notations for potential energy are *PE*, *U*, *V*, and E_p.

Potential energy is the energy by virtue of an object's position relative to other objects. Potential energy is often associated with restoring forces such as a spring or the force of gravity. The action of stretching a spring or lifting a mass is performed by an external force that works against the force field of the potential. This work is stored in the force field, which is said to be stored as potential energy. If the external force is removed the force field acts on the body to perform the work as it moves the body back to the initial position, reducing the stretch of the spring or causing a body to fall.

Consider a ball whose mass is m and whose height is h. The acceleration g of free fall is approximately constant, so the weight force of the ball mg is constant. Force × displacement gives the work done, which is equal to the gravitational potential energy, thus

$$U_g = mgh$$

The more formal definition is that potential energy is the energy difference between the energy of an object in a given position and its energy at a reference position.

Work and Potential Energy

Potential energy is closely linked with forces. If the work done by a force on a body that moves from *A* to *B* does not depend on the path between these points (if the work is done by a conservative force), then the work of this force measured from *A* assigns a scalar value to every other point in space and defines a scalar potential field. In this case, the force can be defined as the negative of the vector gradient of the potential field.

If the work for an applied force is independent of the path, then the work done by the force is evaluated at the start and end of the trajectory of the point of application. This means that there is a function $U(\mathbf{x})$, called a "potential," that can be evaluated at the two points $\mathbf{x}_A$ and $\mathbf{x}_B$ to obtain the work over any trajectory between these two points. It is tradition to define this function with a negative sign so that positive work is a reduction in the potential, that is

$$W = \int_C \mathbf{F} \cdot d\mathbf{x} = U(\mathbf{x}_A) - U(\mathbf{x}_B)$$

where C is the trajectory taken from A to B. Because the work done is independent of the path taken, then this expression is true for any trajectory, C, from A to B.

The function $U(\mathbf{x})$ is called the potential energy associated with the applied force. Examples of forces that have potential energies are gravity and spring forces.

Derivable from a Potential

In this topic the relationship between work and potential energy is presented in more detail. The line integral that defines work along curve C takes a special form if the force $\mathbf{F}$ is related to a scalar field $\varphi(\mathbf{x})$ so that,

$$\mathbf{F} = \nabla \Phi = \left(\frac{\partial \Phi}{\partial x}, \frac{\partial \Phi}{\partial y}, \frac{\partial \Phi}{\partial z} \right).$$

In this case, work along the curve is given by,

$$W = \int_C \mathbf{F} \cdot d\mathbf{x} = \int_C \nabla \Phi \cdot d\mathbf{x},$$

which can be evaluated using the gradient theorem to obtain,

$$W = \Phi(\mathbf{x}_B) - \Phi(\mathbf{x}_A).$$

This shows that when forces are derivable from a scalar field, the work of those forces along a curve C is computed by evaluating the scalar field at the start point A and the end point B of the curve. This means the work integral does not depend on the path between A and B and is said to be independent of the path.

Potential energy $U=-\varphi(\mathbf{x})$ is traditionally defined as the negative of this scalar field so that work by the force field decreases potential energy, that is

$$W = U(\mathbf{x}_A) - U(\mathbf{x}_B).$$

In this case, the application of the del operator to the work function yields,

$$\nabla W = -\nabla U = -\left(\frac{\partial U}{\partial x}, \frac{\partial U}{\partial y}, \frac{\partial U}{\partial z} \right) = \mathbf{F},$$

and the force $\mathbf{F}$ is said to be "derivable from a potential." This also necessarily implies that $\mathbf{F}$ must be a conservative vector field. The potential U defines a force $\mathbf{F}$ at every point $\mathbf{x}$ in space, so the set of forces is called a force field.

Computing Potential Energy

Given a force field **F(x)**, evaluation of the work integral using the gradient theorem can be used to find the scalar function associated with potential energy. This is done by introducing a parameterized curve γ(t)=**r**(t) from γ(a)=A to γ(b)=B, and computing,

$$\int_\gamma \nabla\Phi(\mathbf{r})\cdot d\mathbf{r} = \int_a^b \nabla\Phi(\mathbf{r}(t))\cdot\mathbf{r}'(t)dt,$$

$$= \int_a^b \frac{d}{dt}\Phi(\mathbf{r}(t))dt = \Phi(\mathbf{r}(b)) - \Phi(\mathbf{r}(a)) = \Phi(\mathbf{x}_B) - \Phi(\mathbf{x}_A).$$

For the force field **F**, let **v**= d**r**/dt, then the gradient theorem yields,

$$\int_\gamma \mathbf{F}\cdot d\mathbf{r} = \int_a^b \mathbf{F}\cdot\mathbf{v}dt,$$

$$= -\int_a^b \frac{d}{dt}U(\mathbf{r}(t))dt = U(\mathbf{x}_A) - U(\mathbf{x}_B).$$

The power applied to a body by a force field is obtained from the gradient of the work, or potential, in the direction of the velocity **v** of the point of application, that is

$$P(t) = -\nabla U\cdot\mathbf{v} = \mathbf{F}\cdot\mathbf{v}.$$

Examples of work that can be computed from potential functions are gravity and spring forces.

Potential Energy for Near Earth Gravity

A trebuchet uses the gravitational potential energy of the counterweight to throw projectiles over two hundred meters.

For small height changes, gravitational potential energy can be computed using,

$$U_g = mgh,$$

where m is the mass in kg, g is the local gravitational field (9.8 metres per second squared on earth), h is the height above a reference level in metres, and U is the energy in joules.

In classical physics, gravity exerts a constant downward force $\boldsymbol{F}$=(0, 0, F_z) on the center of mass of a body moving near the surface of the Earth. The work of gravity on a body moving along a trajectory $\boldsymbol{r}(t) = (x(t), y(t), z(t))$, such as the track of a roller coaster is calculated using its velocity, $\boldsymbol{v}=(v_x, v_y, v_z)$, to obtain

$$W = \int_{t_1}^{t_2} \mathbf{F}\cdot\mathbf{v}dt = \int_{t_1}^{t_2} F_z v_z dt = F_z \Delta z.$$

where the integral of the vertical component of velocity is the vertical distance. The work of gravity depends only on the vertical movement of the curve $\boldsymbol{r}(t)$.

Potential Energy for a Linear Spring

Springs are used for storing elastic potential energy.

A horizontal spring exerts a force $\mathbf{F} = (-kx, 0, 0)$ that is proportional to its deformation in the axial or x direction. The work of this spring on a body moving along the space curve $\mathbf{s}(t) = (x(t), y(t), z(t))$, is calculated using its velocity, $\mathbf{v} = (v_x, v_y, v_z)$, to obtain.

$$W = \int_0^t \mathbf{F}\cdot\mathbf{v}dt = -\int_0^t kxv_x dt = -\int_0^t kx\frac{dx}{dt}dt = \int_{x(t_0)}^{x(t)} kx\,dx = \frac{1}{2}kx^2$$

For convenience, consider contact with the spring occurs at $t = 0$, then the integral of the product of the distance x and the x-velocity, xv_x, is $x^2/2$.

The function,

$$U(x) = \frac{1}{2}kx^2,$$

is called the potential energy of a linear spring.

Elastic potential energy is the potential energy of an elastic object (for example a bow or a catapult) that is deformed under tension or compression (or stressed in formal terminology). It arises as a consequence of a force that tries to restore the object to its original shape, which is most often the electromagnetic force between the atoms and molecules that constitute the object. If the stretch is released, the energy is transformed into kinetic energy.

Potential Energy for Gravitational Forces between Two Bodies

The gravitational potential function, also known as gravitational potential energy, is:

$$U = -\frac{GMm}{r},$$

The negative sign follows the convention that work is gained from a loss of potential energy.

Derivation

The gravitational force between two bodies of mass M and m separated by a distance r is given by Newton's law,

$$\mathbf{F} = -\frac{GMm}{r^2}\hat{\mathbf{r}},$$

where $\hat{\mathbf{r}}$ is a vector of length 1 pointing from M to m and G is the gravitational constant.

Let the mass m move at the velocity $\mathbf{v}$ then the work of gravity on this mass as it moves from position $\mathbf{r}(t_1)$ to $\mathbf{r}(t_2)$ is given by,

$$W = -\int_{\mathbf{r}(t_1)}^{\mathbf{r}(t_2)} \frac{GMm}{r^3}\mathbf{r}\cdot d\mathbf{r} = -\int_{t_1}^{t_2} \frac{GMm}{r^3}\mathbf{r}\cdot\mathbf{v}dt.$$

The position and velocity of the mass m are given by,

$$\mathbf{r} = r\mathbf{e}_r, \qquad \mathbf{v} = \dot{r}\mathbf{e}_r + r\dot{\theta}\mathbf{e}_t,$$

where $\mathbf{e}_r$ and $\mathbf{e}_t$ are the radial and tangential unit vectors directed relative to the vector from M to m. Use this to simplify the formula for work of gravity to,

$$W = -\int_{t_1}^{t_2} \frac{GmM}{r^3}(r\mathbf{e}_r)\cdot(\dot{r}\mathbf{e}_r + r\dot{\theta}\mathbf{e}_t)dt = -\int_{t_1}^{t_2} \frac{GmM}{r^3}r\dot{r}dt = \frac{GMm}{r(t_2)} - \frac{GMm}{r(t_1)}.$$

This calculation uses the fact that,

$$\frac{d}{dt}r^{-1} = -r^{-2}\dot{r} = -\frac{\dot{r}}{r^2}.$$

Potential Energy for Electrostatic Forces between Two Bodies

The electrostatic force exerted by a charge Q on another charge q separated by a distance r is given by Coulomb's Law,

$$\mathbf{F} = \frac{1}{4\pi\varepsilon_0}\frac{Qq}{r^2}\hat{\mathbf{r}},$$

where $\hat{\mathbf{r}}$ is a vector of length 1 pointing from Q to q and ε_0 is the vacuum permittivity. This may also be written using Coulomb constant $k_e = 1/4\pi\varepsilon_0$.

The work W required to move q from A to any point B in the electrostatic force field is given by the potential function,

$$U(r) = \frac{1}{4\pi\varepsilon_0}\frac{Qq}{r}.$$

Reference Level

The potential energy is a function of the state a system is in, and is defined relative to that for a particular state. This reference state is not always a real state; it may also be a limit, such as with the distances between all bodies tending to infinity, provided that the energy involved in tending to that limit is finite, such as in the case of inverse-square law forces. Any arbitrary reference state could be used; therefore it can be chosen based on convenience.

Typically the potential energy of a system depends on the *relative* positions of its components only, so the reference state can also be expressed in terms of relative positions.

Gravitational Potential Energy

Gravitational energy is the potential energy associated with gravitational force, as work is required to elevate objects against Earth's gravity. The potential energy due to elevated positions is called gravitational potential energy, and is evidenced by water in an elevated reservoir or kept behind a dam. If an object falls from one point to another point inside a gravitational field, the force of gravity will do positive work on the object, and the gravitational potential energy will decrease by the same amount.

Gravitational force keeps the planets in orbit around the Sun.

Consider a book placed on top of a table. As the book is raised from the floor to the table, some external force works against the gravitational force. If the book falls back to the floor, the "falling" energy the book receives is provided by the gravitational force. Thus, if the book falls off the table, this potential energy goes to accelerate the mass of the book and is converted into kinetic energy. When the book hits the floor this kinetic energy is converted into heat, deformation, and sound by the impact.

The factors that affect an object's gravitational potential energy are its height relative to some reference point, its mass, and the strength of the gravitational field it is in. Thus, a book lying on a table has less gravitational potential energy than the same book on top of a taller cupboard and less gravitational potential energy than a heavier book lying on the same table. An object at a certain height above the Moon's surface has less gravitational potential energy than at the same height above the Earth's surface because the Moon's gravity is weaker. "Height" in the common sense of the term cannot be used for gravitational potential energy calculations when gravity is not assumed to be a constant.

Local Approximation

The strength of a gravitational field varies with location. However, when the change of distance is small in relation to the distances from the center of the source of the gravitational field, this variation in field strength is negligible and we can assume that the force of gravity on a particular object is constant. Near the surface of the Earth, for example, we assume that the acceleration due to gravity is a constant g = 9.8 m/s² ("standard gravity"). In this case, a simple expression for gravitational potential energy can be derived using the $W = Fd$ equation for work, and the equation,

$$W_F = -\Delta U_F.$$

The amount of gravitational potential energy held by an elevated object is equal to the work done against gravity in lifting it. The work done equals the force required to move it upward multiplied with the vertical distance it is moved (remember $W = Fd$). The upward force required while moving at a constant velocity is equal to the weight, mg, of an object, so the work done in lifting it through a height h is the product mgh. Thus, when accounting only for mass, gravity, and altitude, the equation is:

$$U = mgh$$

where U is the potential energy of the object relative to its being on the Earth's surface, m is the mass of the object, g is the acceleration due to gravity, and h is the altitude of the object. If m is expressed in kilograms, g in m/s² and h in metres then U will be calculated in joules.

Hence, the potential difference is,

$$\Delta U = mg\Delta h.$$

General Formula

However, over large variations in distance, the approximation that g is constant is no longer valid, and we have to use calculus and the general mathematical definition of work to determine gravitational potential energy. For the computation of the potential energy, we can integrate the gravitational force, whose magnitude is given by Newton's law of gravitation, with respect to the distance r between the two bodies. Using that definition, the gravitational potential energy of a system of masses m_1 and M_2 at a distance r using gravitational constant G is,

$$U = -G\frac{m_1 M_2}{r} + K,$$

where K is an arbitrary constant dependent on the choice of datum from which potential is measured. Choosing the convention that K=0 (i.e. in relation to a point at infinity) makes calculations simpler, albeit at the cost of making U negative; for why this is physically reasonable.

Given this formula for U, the total potential energy of a system of n bodies is found by summing, for all $\frac{n(n-1)}{2}$ pairs of two bodies, the potential energy of the system of those two bodies.

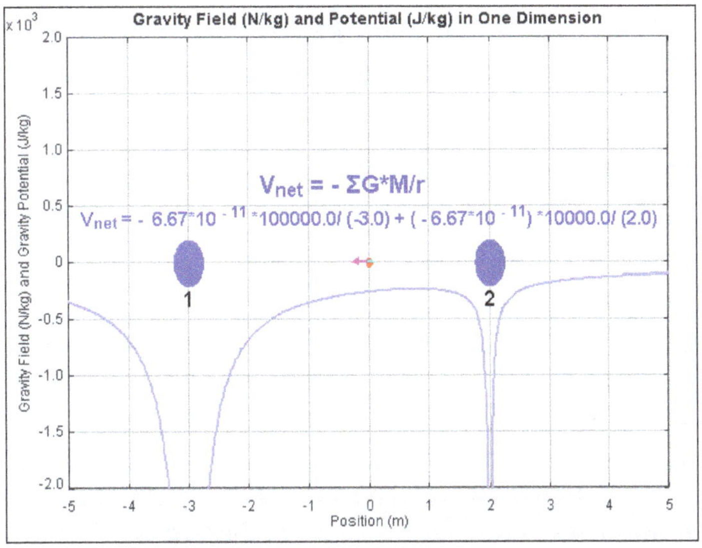

Gravitational potential summation.

Considering the system of bodies as the combined set of small particles the bodies consist of, and applying the previous on the particle level we get the negative gravitational binding energy. This potential energy is more strongly negative than the total potential energy of the system of bodies as such since it also includes the negative gravitational binding energy of each body. The potential energy of the system of bodies as such is the negative of the energy needed to separate the bodies from each other to infinity, while the gravitational binding energy is the energy needed to separate all particles from each other to infinity.

$$U = -m\left(G\frac{M_1}{r_1} + G\frac{M_2}{r_2}\right)$$

therefore,

$$U = -m\sum G\frac{M}{r}.$$

Negative Gravitational Energy

As with all potential energies, only differences in gravitational potential energy matter for most physical purposes, and the choice of zero point is arbitrary. Given that there is no reasonable criterion for preferring one particular finite *r* over another, there seem to be only two reasonable choices for the distance at which *U* becomes zero: $r = 0$ and $r = \infty$. The choice of $U = 0$ at infinity may seem peculiar, and the consequence that gravitational energy is always negative may seem counterintuitive, but this choice allows gravitational potential energy values to be finite, albeit negative.

The singularity at $r = 0$ in the formula for gravitational potential energy means that the only other apparently reasonable alternative choice of convention, with $U = 0$ for $r = 0$, would result in potential energy being positive, but infinitely large for all nonzero values of *r*, and would make calculations involving sums or differences of potential energies beyond what is possible with the real number system. Since physicists abhor infinities in their calculations, and *r* is always non-zero in practice, the choice of $U = 0$ at infinity is by far the more preferable choice, even if the idea of negative energy in a gravity well appears to be peculiar at first.

The negative value for gravitational energy also has deeper implications that make it seem more reasonable in cosmological calculations where the total energy of the universe can meaningfully be considered.

Uses

Gravitational potential energy has a number of practical uses, notably the generation of pumped-storage hydroelectricity. For example, in Dinorwig, Wales, there are two lakes, one at a higher elevation than the other. At times when surplus electricity is not required (and so is comparatively cheap), water is pumped up to the higher lake, thus converting the electrical energy (running the pump) to gravitational potential energy. At times of peak demand for electricity, the water flows back down through electrical generator turbines, converting the potential energy into kinetic energy and then back into electricity. The process is not completely efficient and some of the original energy from the surplus electricity is in fact lost to friction.

Gravitational potential energy is also used to power clocks in which falling weights operate the mechanism.

It's also used by counterweights for lifting up an elevator, crane, or sash window. Roller coasters are an entertaining way to utilize potential energy – chains are used to move a car up an incline (building up gravitational potential energy), to then have that energy converted into kinetic energy as it falls.

Another practical use is utilizing gravitational potential energy to descend (perhaps coast) downhill in transportation such as the descent of an automobile, truck, railroad train, bicycle, airplane, or fluid in a pipeline. In some cases the kinetic energy obtained from the potential energy of descent may be used to start ascending the next grade such as what happens when a road is undulating and has frequent dips. The commercialization of stored energy (in the form of rail cars raised to higher elevations) that is then converted to electrical energy when needed by an electrical grid, is being undertaken in the United States in a system called Advanced Rail Energy Storage (ARES).

Forces and Potential Energy

Potential energy is closely linked with forces. If the work done by a force on a body that moves from *A* to *B* does not depend on the path between these points, then the work of this force measured from *A* assigns a scalar value to every other point in space and defines a scalar potential field. In this case, the force can be defined as the negative of the vector gradient of the potential field.

For example, gravity is a conservative force. The associated potential is the gravitational potential, often denoted by ϕ or V, corresponding to the energy per unit mass as a function of position. The gravitational potential energy of two particles of mass *M* and *m* separated by a distance *r* is,

$$U = -\frac{GMm}{r},$$

The gravitational potential (specific energy) of the two bodies is,

$$\phi = -\left(\frac{GM}{r} + \frac{Gm}{r}\right) = -\frac{G(M+m)}{r} = -\frac{GMm}{\mu r} = \frac{U}{\mu}.$$

where μ is the reduced mass.

The work done against gravity by moving an infinitesimal mass from point A with $U = a$ to point B with $U = b$ is $(b - a)$ and the work done going back the other way is $(a - b)$ so that the total work done in moving from A to B and returning to A is,

$$U_{A\to B\to A} = (b-a)+(a-b) = 0.$$

If the potential is redefined at A to be $a + c$ and the potential at B to be $b + c$, where c is a constant (i.e. c can be any number, positive or negative, but it must be the same at A as it is at B) then the work done going from A to B is,

$$U_{A\to B} = (b+c)-(a+c) = b-a$$

as before.

In practical terms, this means that one can set the zero of U and ϕ anywhere one likes. One may set it to be zero at the surface of the Earth, or may find it more convenient to set zero at infinity.

A conservative force can be expressed in the language of differential geometry as a closed form. As Euclidean space is contractible, its de Rham cohomology vanishes, so every closed form is also an exact form, and can be expressed as the gradient of a scalar field. This gives a mathematical justification of the fact that all conservative forces are gradients of a potential field.

Power

In physics, power is the rate of doing work or of transferring heat, i.e. the amount of energy transferred or converted per unit time. Having no direction, it is a scalar quantity. In the International System of Units, the unit of power is the joule per second (J/s), known as the watt (W) in honour of James Watt, the eighteenth-century developer of the condenser steam engine. Another common and traditional measure is horsepower (comparing to the power of a horse); 1 horsepower equals about 745.7 watts. Being the rate of work, the equation for power can be written as:

$$\text{power} = \frac{\text{work}}{\text{time}}$$

As a physical concept, power requires both a change in the physical system and a specified time in which the change occurs. This is distinct from the concept of work, which is measured only in terms of a net change in the state of the physical system. The same amount of work is done when carrying a load up a flight of stairs whether the person carrying it walks or runs, but more power is needed for running because the work is done in a shorter amount of time.

The output power of an electric motor is the product of the torque that the motor generates and the angular velocity of its output shaft. The power involved in moving a ground vehicle is the product of the traction force on the wheels and the velocity of the vehicle. The power of a jet-propelled vehicle is the product of the engine thrust and the velocity of the vehicle. The rate at which a light bulb converts electrical energy into light and heat is measured in watts—the higher the wattage, the more power, or equivalently the more electrical energy is used per unit time.

Units

The dimension of power is energy divided by time. The SI unit of power is the watt (W), which is equal to one joule per second. Other units of power include ergs per second (erg/s), horsepower (hp), metric horsepower (Pferdestärke (PS) or cheval vapeur (CV)), and foot-pounds per minute. One horsepower is equivalent to 33,000 foot-pounds per minute, or the power required to lift 550 pounds by one foot in one second, and is equivalent to about 746 watts. Other units include dBm, a logarithmic measure relative to a reference of 1 milliwatt; food calories per hour (often referred to as kilocalories per hour); BTU per hour (BTU/h); and tons of refrigeration (12,000 BTU/h).

Equations for Power

Power, as a function of time, is the rate (i.e. *derivative*) at which work is done, so can be expressed by this equation:

$$P = \frac{dW}{dt}$$

where P is power, W is work, and t is time. Because work is a force $\mathbf{F}$ applied over a distance $\mathbf{x}$,

$$W = \mathbf{F} \cdot \mathbf{x}$$

for a constant force, power can be rewritten as:

$$P = \frac{dW}{dt} = \frac{d}{dt}(\mathbf{F} \cdot \mathbf{x}) = \mathbf{F} \cdot \frac{d\mathbf{x}}{dt} = \mathbf{F} \cdot \mathbf{v}$$

In fact, this is valid for *any* force, as a consequence of applying the fundamental theorem of calculus.

Average Power

As a simple example, burning one kilogram of coal releases much more energy than does detonating a kilogram of TNT, but because the TNT reaction releases energy much more quickly, it delivers far more power than the coal. If ΔW is the amount of work performed during a period of time of duration Δt, the average power P_{avg} over that period is given by the formula,

$$P_{avg} = \frac{\Delta W}{\Delta t}.$$

It is the average amount of work done or energy converted per unit of time. The average power is often simply called "power" when the context makes it clear.

The instantaneous power is then the limiting value of the average power as the time interval Δt approaches zero.

$$P = \lim_{\Delta t \to 0} P_{avg} = \lim_{\Delta t \to 0} \frac{\Delta W}{\Delta t} = \frac{dW}{dt}.$$

In the case of constant power P, the amount of work performed during a period of duration t is given by:

$$W = Pt.$$

In the context of energy conversion, it is more customary to use the symbol E rather than W.

Mechanical Power

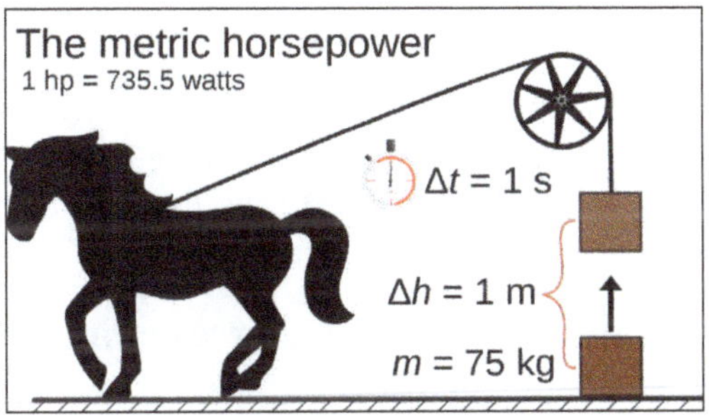

One metric horsepower is needed to lift 75 kilograms by 1 meter in 1 second.

Power in mechanical systems is the combination of forces and movement. In particular, power is the product of a force on an object and the object's velocity, or the product of a torque on a shaft and the shaft's angular velocity.

Mechanical power is also described as the time derivative of work. In mechanics, the work done by a force **F** on an object that travels along a curve C is given by the line integral:

$$W_C = \int_C \mathbf{F} \cdot \mathbf{v} \, dt = \int_C \mathbf{F} \cdot d\mathbf{x},$$

where **x** defines the path C and **v** is the velocity along this path.

If the force **F** is derivable from a potential (conservative), then applying the gradient theorem (and remembering that force is the negative of the gradient of the potential energy) yields:

$$W_C = U(B) - U(A),$$

where A and B are the beginning and end of the path along which the work was done.

The power at any point along the curve C is the time derivative,

$$P(t) = \frac{dW}{dt} = \mathbf{F} \cdot \mathbf{v} = -\frac{dU}{dt}.$$

In one dimension, this can be simplified to:

$$P(t) = F \cdot v.$$

In rotational systems, power is the product of the torque τ and angular velocity ω,

$$P(t) = \tau \cdot \omega,$$

where $\boldsymbol{\omega}$ measured in radians per second. The $\cdot$ represents scalar product.

In fluid power systems such as hydraulic actuators, power is given by,

$$P(t) = pQ,$$

where p is pressure in pascals, or N/m² and Q is volumetric flow rate in m³/s in SI units.

Mechanical Advantage

If a mechanical system has no losses, then the input power must equal the output power. This provides a simple formula for the mechanical advantage of the system.

Let the input power to a device be a force F_A acting on a point that moves with velocity v_A and the output power be a force F_B acts on a point that moves with velocity v_B. If there are no losses in the system, then

$$P = F_B v_B = F_A v_A,$$

and the mechanical advantage of the system (output force per input force) is given by,

$$\mathrm{MA} = \frac{F_B}{F_A} = \frac{v_A}{v_B}.$$

The similar relationship is obtained for rotating systems, where T_A and ω_A are the torque and angular velocity of the input and T_B and ω_B are the torque and angular velocity of the output. If there are no losses in the system, then

$$P = T_A \omega_A = T_B \omega_B,$$

which yields the mechanical advantage,

$$\mathrm{MA} = \frac{T_B}{T_A} = \frac{\omega_A}{\omega_B}.$$

These relations are important because they define the maximum performance of a device in terms of velocity ratios determined by its physical dimensions.

Impulse

In classical mechanics, impulse (symbolized by J or Imp) is the integral of a force, F, over the time interval, t, for which it acts. Since force is a vector quantity, impulse is also a vector in the same direction. Impulse applied to an object produces an equivalent vector change in its linear momentum, also in the same direction. The SI unit of impulse is the newton second (N·s), and the dimensionally equivalent unit of momentum is the kilogram meter per second (kg·m/s). The corresponding English engineering units are the pound-second (lbf·s) and the slug-foot per second (slug·ft/s).

A resultant force causes acceleration and a change in the velocity of the body for as long as it acts. A resultant force applied over a longer time therefore produces a bigger change in linear momentum than the same force applied briefly: the change in momentum is equal to the product of the average force and duration. Conversely, a small force applied for a long time produces the same change in momentum—the same impulse—as a larger force applied briefly.

$$J = F_{\text{average}}(t_2 - t_1)$$

The impulse is the integral of the resultant force (F) with respect to time:

$$J = \int F \, \mathrm{d}t$$

Mathematical Derivation in the Case of an Object of Constant Mass

Impulse **J** produced from time t_1 to t_2 is defined to be,

$$\mathbf{J} = \int_{t_1}^{t_2} \mathbf{F} \, \mathrm{d}t$$

where **F** is the resultant force applied from t_1 to t_2.

From Newton's second law, force is related to momentum **p** by,

$$\mathbf{F} = \frac{\mathrm{d}\mathbf{p}}{\mathrm{d}t}$$

Therefore,

$$\begin{aligned} \mathbf{J} &= \int_{t_1}^{t_2} \frac{\mathrm{d}\mathbf{p}}{\mathrm{d}t} \mathrm{d}t \\ &= \int_{\mathbf{p}_1}^{\mathbf{p}_2} \mathrm{d}\mathbf{p} \\ &= \mathbf{p}_2 - \mathbf{p}_1 = \Delta\mathbf{p} \end{aligned}$$

where $\Delta\mathbf{p}$ is the change in linear momentum from time t_1 to t_2. This is often called the impulse-momentum theorem (analogous to the work-energy theorem).

A large force applied for a very short duration, such as a golf shot, is often described as the club giving the ball an impulse.

As a result, an impulse may also be regarded as the change in momentum of an object to which a resultant force is applied. The impulse may be expressed in a simpler form when the mass is constant:

$$\mathbf{J} = \int_{t_1}^{t_2} \mathbf{F}\, dt = \Delta\mathbf{p} = m\mathbf{v}_2 - m\mathbf{v}_1$$

where

F is the resultant force applied,

t_1 and t_2 are times when the impulse begins and ends, respectively,

m is the mass of the object,

$\mathbf{v}_2$ is the final velocity of the object at the end of the time interval, and

$\mathbf{v}_1$ is the initial velocity of the object when the time interval begins.

Impulse has the same units and dimensions (MLT^{-1}) as momentum. In the International System of Units, these are kg·m/s = N·s. In English engineering units, they are slug·ft/s = lbf·s.

The term "impulse" is also used to refer to a fast-acting force or impact. This type of impulse is often *idealized* so that the change in momentum produced by the force happens with no change in time. This sort of change is a step change, and is not physically possible. However, this is a useful model for computing the effects of ideal collisions (such as in game physics engines). Additionally, in rocketry, the term "total impulse" is commonly used and is considered synonymous with the term "impulse".

Variable Mass

The application of Newton's second law for variable mass allows impulse and momentum to be used as analysis tools for jet- or rocket-propelled vehicles. In the case of rockets, the impulse imparted can be normalized by unit of propellant expended, to create a performance parameter, specific impulse. This fact can be used to derive the Tsiolkovsky rocket equation, which relates the vehicle's propulsive change in velocity to the engine's specific impulse (or nozzle exhaust velocity) and the vehicle's propellant-mass ratio.

Momentum

In Newtonian mechanics, linear momentum, translational momentum, or simply momentum (pl. momenta) is the product of the mass and velocity of an object. It is a vector quantity, possessing a magnitude and a direction in three-dimensional space. If m is an object's mass and **v** is the velocity (also a vector quantity), then the momentum is,

$$\mathbf{p} = m\mathbf{v},$$

In SI units, it is measured in kilogram meters per second (kg·m/s). Newton's second law of motion states that a body's rate of change in momentum is equal to the net force acting on it.

Momentum depends on the frame of reference, but in any inertial frame it is a *conserved* quantity, meaning that if a closed system is not affected by external forces, its total linear momentum does not change. Momentum is also conserved in special relativity (with a modified formula) and, in a modified form, in electrodynamics, quantum mechanics, quantum field theory, and general relativity. It is an expression of one of the fundamental symmetries of space and time: translational symmetry.

Advanced formulations of classical mechanics, Lagrangian and Hamiltonian mechanics, allow one to choose coordinate systems that incorporate symmetries and constraints. In these systems the conserved quantity is generalized momentum, and in general this is different from the kinetic momentum defined above. The concept of generalized momentum is carried over into quantum mechanics, where it becomes an operator on a wave function. The momentum and position operators are related by the Heisenberg uncertainty principle.

In continuous systems such as electromagnetic fields, fluids and deformable bodies, a momentum density can be defined, and a continuum version of the conservation of momentum leads to equations such as the Navier–Stokes equations for fluids or the Cauchy momentum equation for deformable solids or fluids.

Newtonian

Momentum is a vector quantity: it has both magnitude and direction. Since momentum has a direction, it can be used to predict the resulting direction and speed of motion of objects after they collide. Below, the basic properties of momentum are described in one dimension. The vector equations are almost identical to the scalar equations.

Single Particle

The momentum of a particle is conventionally represented by the letter p. It is the product of two quantities, the particle's mass (represented by the letter m) and its velocity (v):

$$p = mv.$$

The unit of momentum is the product of the units of mass and velocity. In SI units, if the mass is in kilograms and the velocity is in meters per second then the momentum is in kilogram meters per second (kg·m/s). In cgs units, if the mass is in grams and the velocity in centimeters per second, then the momentum is in gram centimeters per second (g·cm/s).

Being a vector, momentum has magnitude and direction. For example, a 1 kg model airplane, traveling due north at 1 m/s in straight and level flight, has a momentum of 1 kg·m/s due north measured with reference to the ground.

Many Particles

The momentum of a system of particles is the vector sum of their momenta. If two particles have respective masses m_1 and m_2, and velocities v_1 and v_2, the total momentum is,

$$\begin{aligned} p &= p_1 + p_2 \\ &= m_1 v_1 + m_2 v_2. \end{aligned}$$

The momenta of more than two particles can be added more generally with the following:

$$p = \sum_i m_i v_i.$$

A system of particles has a center of mass, a point determined by the weighted sum of their positions:

$$r_{\text{cm}} = \frac{m_1 r_1 + m_2 r_2 + \cdots}{m_1 + m_2 + \cdots} = \frac{\sum_i m_i r_i}{\sum_i m_i}.$$

If one or more of the particles is moving, the center of mass of the system will generally be moving as well (unless the system is in pure rotation around it). If the total mass of the particles is m, and the center of mass is moving at velocity v_{cm}, the momentum of the system is:

$$p = m v_{\text{cm}}.$$

This is known as Euler's first law.

Relation to Force

If the net force F applied to a particle is constant, and is applied for a time interval Δt, the momentum of the particle changes by an amount,

$$\Delta p = F \Delta t.$$

In differential form, this is Newton's second law; the rate of change of the momentum of a particle is equal to the instantaneous force F acting on it,

$$F = \frac{dp}{dt}.$$

If the net force experienced by a particle changes as a function of time, $F(t)$, the change in momentum (or impulse J) between times t_1 and t_2 is,

$$\Delta p = J = \int_{t_1}^{t_2} F(t)\,dt.$$

Impulse is measured in the derived units of the newton second (1 N·s = 1 kg·m/s) or dyne second (1 dyne·s = 1 g·cm/s).

Under the assumption of constant mass m, it is equivalent to write,

$$F = \frac{d(mv)}{dt} = m\frac{dv}{dt} = ma,$$

hence the net force is equal to the mass of the particle times its acceleration.

Example: A model airplane of mass 1 kg accelerates from rest to a velocity of 6 m/s due north in 2 s. The net force required to produce this acceleration is 3 newtons due north. The change in momentum is 6 kg·m/s due north. The rate of change of momentum is 3 (kg·m/s)/s due north which is numerically equivalent to 3 newtons.

Conservation

In a closed system (one that does not exchange any matter with its surroundings and is not acted on by external forces) the total momentum is constant. This fact, known as the law of conservation of momentum, is implied by Newton's laws of motion. Suppose, for example, that two particles interact. Because of the third law, the forces between them are equal and opposite. If the particles are numbered 1 and 2, the second law states that $F_1 = dp_{1/dt}$ and $F_2 = dp_{2/dt}$. Therefore,

$$\frac{dp_1}{dt} = -\frac{dp_2}{dt},$$

with the negative sign indicating that the forces oppose. Equivalently,

$$\frac{d}{dt}(p_1 + p_2) = 0.$$

If the velocities of the particles are u_1 and u_2 before the interaction, and afterwards they are v_1 and v_2, then

$$m_1u_1 + m_2u_2 = m_1v_1 + m_2v_2.$$

This law holds no matter how complicated the force is between particles. Similarly, if there are several particles, the momentum exchanged between each pair of particles adds up to zero, so the total change in momentum is zero. This conservation law applies to all interactions, including collisions and separations caused by explosive forces. It can also be generalized to situations where Newton's laws do not hold, for example in the theory of relativity and in electrodynamics.

Dependence on Reference Frame

Momentum is a measurable quantity, and the measurement depends on the motion of the observer. For example: if an apple is sitting in a glass elevator that is descending, an outside observer, looking into the elevator, sees the apple moving, so, to that observer, the apple has a non-zero momentum. To someone inside the elevator, the apple does not move, so, it has zero momentum. The two observers each have a frame of reference, in which, they observe motions, and, if the elevator is descending steadily, they will see behavior that is consistent with those same physical laws.

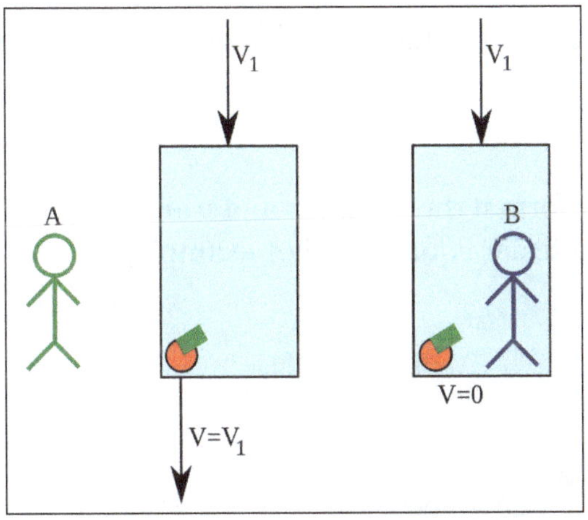

Newton's apple in Einstein's elevator. In person A's frame of reference, the apple has non-zero velocity and momentum. In the elevator's and person B's frames of reference, it has zero velocity and momentum.

Suppose a particle has position x in a stationary frame of reference. From the point of view of another frame of reference, moving at a uniform speed u, the position (represented by a primed coordinate) changes with time as,

$$x' = x - ut.$$

This is called a Galilean transformation. If the particle is moving at speed $dx/dt = v$ in the first frame of reference, in the second, it is moving at speed,

$$v' = \frac{dx'}{dt} = v - u.$$

Since u does not change, the accelerations are the same:

$$a' = \frac{dv'}{dt} = a.$$

Thus, momentum is conserved in both reference frames. Moreover, as long as the force has the same form, in both frames, Newton's second law is unchanged. Forces such as Newtonian gravity, which depend only on the scalar distance between objects, satisfy this criterion. This independence of reference frame is called Newtonian relativity or Galilean invariance.

A change of reference frame, can, often, simplify calculations of motion. For example, in a collision of two particles, a reference frame can be chosen, where, one particle begins at rest. Another, commonly used reference frame, is the center of mass frame – one that is moving with the center of mass. In this frame, the total momentum is zero.

Multiple Dimensions

Real motion has both direction and velocity and must be represented by a vector. In a coordinate system with x, y, z axes, velocity has components v_x in the x-direction, v_y in the y-direction, v_z in the z-direction. The vector is represented by a boldface symbol:

$$\mathbf{v} = (v_x, v_y, v_z).$$

Similarly, the momentum is a vector quantity and is represented by a boldface symbol:

$$\mathbf{p} = (p_x, p_y, p_z).$$

The equations work in vector form if the scalars p and v are replaced by vectors $\mathbf{p}$ and $\mathbf{v}$. Each vector equation represents three scalar equations. For example,

$$\mathbf{p} = m\mathbf{v}$$

represents three equations:

$$p_x = mv_x$$
$$p_y = mv_y$$
$$p_z = mv_z.$$

The kinetic energy equations are exceptions to the above replacement rule. The equations are still one-dimensional, but each scalar represents the magnitude of the vector, for example,

$$v^2 = v_x^2 + v_y^2 + v_z^2.$$

Each vector equation represents three scalar equations. Often coordinates can be chosen so that only two components are needed, as in the figure. Each component can be obtained separately and the results combined to produce a vector result.

A simple construction involving the center of mass frame can be used to show that if a stationary elastic sphere is struck by a moving sphere, the two will head off at right angles after the collision (as in the figure).

Objects of Variable Mass

The concept of momentum plays a fundamental role in explaining the behavior of variable-mass objects such as a rocket ejecting fuel or a star accreting gas. In analyzing such an object, one treats the object's mass as a function that varies with time: $m(t)$. The momentum of the object at time t is therefore $p(t) = m(t)v(t)$. One might then try to invoke Newton's second law of motion by saying that the external force F on the object is related to its momentum $p(t)$ by $F = dp/dt$, but this is incorrect, as is the related expression found by applying the product rule to $d(mv)/dt$:

$$F = m(t)\frac{dv}{dt} + v(t)\frac{dm}{dt}. \text{ (incorrect)}$$

This equation does not correctly describe the motion of variable-mass objects. The correct equation is,

$$F = m(t)\frac{dv}{dt} - u\frac{dm}{dt},$$

where u is the velocity of the ejected/accreted mass *as seen in the object's rest frame*. This is distinct from v, which is the velocity of the object itself as seen in an inertial frame.

This equation is derived by keeping track of both the momentum of the object as well as the momentum of the ejected/accreted mass (dm). When considered together, the object and the mass (dm) constitute a closed system in which total momentum is conserved.

$$P(t+dt) = (m-dm)(v+dv) + dm(v-u) = mv + mdv - udm = P(t) + mdv - udm$$

Relativistic

Lorentz Invariance

Newtonian physics assumes that absolute time and space exist outside of any observer; this gives rise to Galilean invariance. It also results in a prediction that the speed of light can vary from one reference frame to another. This is contrary to observation. In the special theory of relativity, Einstein keeps the postulate that the equations of motion do not depend on the reference frame, but assumes that the speed of light c is invariant. As a result, position and time in two reference frames are related by the Lorentz transformation instead of the Galilean transformation.

Consider, for example, a reference frame moving relative to another at velocity v in the x direction. The Galilean transformation gives the coordinates of the moving frame as,

$$t' = t$$
$$x' = x - vt$$

while the Lorentz transformation gives ,

$$t' = \gamma\left(t - \frac{vx}{c^2}\right)$$
$$x' = \gamma(x - vt)$$

where γ is the Lorentz factor:

$$\gamma = \frac{1}{\sqrt{1 - v^2 / c^2}}.$$

Newton's second law, with mass fixed, is not invariant under a Lorentz transformation. However, it can be made invariant by making the *inertial mass m* of an object a function of velocity:

$$m = \gamma m_0;$$

m_0 is the object's invariant mass.

The modified momentum,

$$\mathbf{p} = \gamma m_0 \mathbf{v},$$

obeys Newton's second law:

$$\mathbf{F} = \frac{d\mathbf{p}}{dt}.$$

Within the domain of classical mechanics, relativistic momentum closely approximates Newtonian momentum: at low velocity, $\gamma m_0\mathbf{v}$ is approximately equal to $m_0\mathbf{v}$, the Newtonian expression for momentum.

Four-vector Formulation

In the theory of special relativity, physical quantities are expressed in terms of four-vectors that include time as a fourth coordinate along with the three space coordinates. These vectors are generally represented by capital letters, for example **R** for position. The expression for the *four-momentum* depends on how the coordinates are expressed. Time may be given in its normal units or multiplied by the speed of light so that all the components of the four-vector have dimensions of length. If the latter scaling is used, an interval of proper time, τ, defined by,

$$c^2 d\tau^2 = c^2 dt^2 - dx^2 - dy^2 - dz^2,$$

is invariant under Lorentz transformations (in this expression and in what follows the (+ − − −) metric signature has been used,). Mathematically this invariance can be ensured in one of two ways: by treating the four-vectors as Euclidean vectors and multiplying time by $\sqrt{-1}$; or by keeping time a real quantity and embedding the vectors in a Minkowski space. In a Minkowski space, the scalar product of two four-vectors $\mathbf{U} = (U_0,U_1,U_2,U_3)$ and $\mathbf{V} = (V_0,V_1,V_2,V_3)$ is defined as,

$$\mathbf{U}\cdot\mathbf{V} = U_0V_0 - U_1V_1 - U_2V_2 - U_3V_3.$$

In all the coordinate systems, the (contravariant) relativistic four-velocity is defined by,

$$\mathbf{U} \equiv \frac{d\mathbf{R}}{d\tau} = \gamma\frac{d\mathbf{R}}{dt},$$

and the (contravariant) four-momentum is,

$$\mathbf{P} = m_0\mathbf{U},$$

where m_0 is the invariant mass. If $\mathbf{R} = (ct,x,y,z)$ (in Minkowski space), then

$$\mathbf{P} = \gamma m_0\left(c,\mathbf{v}\right) = (mc,\mathbf{p}).$$

Using Einstein's mass-energy equivalence, $E = mc^2$, this can be rewritten as,

$$\mathbf{P} = \left(\frac{E}{c},\mathbf{p}\right).$$

Thus, conservation of four-momentum is Lorentz-invariant and implies conservation of both mass and energy.

The magnitude of the momentum four-vector is equal to m_0c:

$$\|\mathbf{P}\|^2 = \mathbf{P}\cdot\mathbf{P} = \gamma^2 m_0^2(c^2 - v^2) = (m_0c)^2,$$

and is invariant across all reference frames.

The relativistic energy–momentum relationship holds even for massless particles such as photons; by setting $m_0 = 0$ it follows that,

$$E = pc.$$

In a game of relativistic "billiards", if a stationary particle is hit by a moving particle in an elastic collision, the paths formed by the two afterwards will form an acute angle. This is unlike the non-relativistic case where they travel at right angles.

The four-momentum of a planar wave can be related to a wave four-vector ,

$$\mathbf{P} = \left(\frac{E}{c},\vec{\mathbf{p}}\right) = \hbar\mathbf{K} = \hbar\left(\frac{\omega}{c},\vec{\mathbf{k}}\right).$$

For a particle, the relationship between temporal components, $E = \hbar\,\omega$, is the Planck–Einstein relation, and the relation between spatial components, $\mathbf{p} = \hbar\,\mathbf{k}$, describes a de Broglie matter wave.

Generalized

Newton's laws can be difficult to apply to many kinds of motion because the motion is limited by constraints. For example, a bead on an abacus is constrained to move along its wire and a pendulum bob is constrained to swing at a fixed distance from the pivot. Many such constraints can be incorporated by changing the normal Cartesian coordinates to a set of generalized coordinates that may be fewer in number. Refined mathematical methods have been developed for solving mechanics problems in generalized coordinates. They introduce a generalized momentum, also known as the canonical or conjugate momentum, that extends the concepts of both linear momentum and angular momentum. To distinguish it from generalized momentum, the product of mass and velocity is also referred to as mechanical, kinetic or kinematic momentum. The two main methods are described below.

Lagrangian Mechanics

In Lagrangian mechanics, a Lagrangian is defined as the difference between the kinetic energy T and the potential energy V:

$$\mathcal{L} = T - V.$$

If the generalized coordinates are represented as a vector $\mathbf{q} = (q_1, q_2, \ldots, q_N)$ and time differentiation is represented by a dot over the variable, then the equations of motion (known as the Lagrange or Euler–Lagrange equations) are a set of N equations:

$$\frac{d}{dt}\left(\frac{\partial \mathcal{L}}{\partial \dot{q}_j}\right) - \frac{\partial \mathcal{L}}{\partial q_j} = 0.$$

If a coordinate q_i is not a Cartesian coordinate, the associated generalized momentum component p_i does not necessarily have the dimensions of linear momentum. Even if q_i is a Cartesian coordinate, p_i will not be the same as the mechanical momentum if the potential depends on velocity. Some sources represent the kinematic momentum by the symbol $\mathbf{\Pi}$.

In this mathematical framework, a generalized momentum is associated with the generalized coordinates. Its components are defined as,

$$p_j = \frac{\partial \mathcal{L}}{\partial \dot{q}_j}.$$

Each component p_j is said to be the *conjugate momentum* for the coordinate q_j.

Now if a given coordinate q_i does not appear in the Lagrangian (although its time derivative might appear), then

$$p_j = \text{constant}.$$

This is the generalization of the conservation of momentum.

Even if the generalized coordinates are just the ordinary spatial coordinates, the conjugate momenta are not necessarily the ordinary momentum coordinates.

Hamiltonian Mechanics

In Hamiltonian mechanics, the Lagrangian (a function of generalized coordinates and their derivatives) is replaced by a Hamiltonian that is a function of generalized coordinates and momentum. The Hamiltonian is defined as,

$$\mathcal{H}(\mathbf{q},\mathbf{p},t)=\mathbf{p}\cdot\dot{\mathbf{q}}-\mathcal{L}\left(\mathbf{q},\dot{\mathbf{q}},t\right),$$

where the momentum is obtained by differentiating the Lagrangian as above. The Hamiltonian equations of motion are,

$$\dot{q}_i=\frac{\partial\mathcal{H}}{\partial p_i}$$

$$-\dot{p}_i=\frac{\partial\mathcal{H}}{\partial q_i}$$

$$-\frac{\partial\mathcal{L}}{\partial t}=\frac{d\mathcal{H}}{dt}.$$

As in Lagrangian mechanics, if a generalized coordinate does not appear in the Hamiltonian, its conjugate momentum component is conserved.

Symmetry and Conservation

Conservation of momentum is a mathematical consequence of the homogeneity (shift symmetry) of space (position in space is the canonical conjugate quantity to momentum). That is, conservation of momentum is a consequence of the fact that the laws of physics do not depend on position; this is a special case of Noether's theorem.

Collision

In physics, collision (also called impact) is the sudden, forceful coming together in direct contact of two bodies, such as, for example, two billiard balls, a golf club and a ball, a hammer and a nail head, two railroad cars when being coupled together, or a falling object and a floor. Apart from the properties of the materials of the two objects, two factors affect the result of impact: the force and the time during which the objects are in contact. It is a matter of common experience that a hard steel ball dropped on a steel plate will rebound to almost the position from which it was dropped, whereas with a ball of putty or lead there is no rebound. The impact between the steel ball and plate is said to be elastic, and that between the putty or lead balls and plate is inelastic, or plastic; between these extremes there are varying degrees of elasticity and corresponding responses to impact. In a perfectly elastic impact (attained only at the atomic level), none of the kinetic energy of the coacting bodies is lost; in a perfectly plastic impact, the loss of kinetic energy is at a maximum.

In all of the examples of colliding bodies here referred to, the time of contact is extremely short and the force of contact extremely large. It can be shown that, in the limiting case of an "infinite" force acting for an "infinitesimal" time, there is an instantaneous change in the velocity of a body but no change in its position during the period of contact. Forces of this nature are known as impulsive forces and, being difficult to measure or estimate, their effects are measured by the change in the momentum (mass times velocity) of the body. The ballistic pendulum is a device based on this principle.

When two bodies collide, the sum of the momenta of the bodies before impact is equal to the sum of the momenta after impact. The relation between the kinetic energies before and after impact depends, as previously noted, on the elasticity of the bodies. Knowing the initial velocities, the final velocities can be obtained by the simultaneous solution of the momentum and energy equations in the case of perfectly elastic collisions.

Types of Collision

Generally, the law of conservation of momentum holds true in the collision of two masses but there may be some collisions in which Kinetic Energy is not conserved. Depending on the energy conservation, conservation may be of two types:

- Elastic collision: In the elastic collision total momentum, the total energy and the total kinetic energy are conserved. However, the total mechanical energy is not converted into any other energy form as the forces involved in the short interaction are conserved in nature. Consider from the above graph two masses, m_1 and m_2 moving with speed u_1 and u_2. The speed after the collision of these masses is v_1 *and* v_2. The law of conservation of momentum will give:

  $$m_1u_1 + m_2u_2 = m_1v_1 + m_2v_2$$

  The conservation of Kinetic Energy says:

  $$1/2\, m_1u^2_1 + 1/2\, m_2u^2_2 = 1/2\, m_1v^2_1 + 1/2\, m_2v^2_2$$

- Inelastic collision: In the inelastic collision, the objects stick to each other or move in the same direction. The total kinetic energy in this form of collision is not conserved but the total momentum and energy are conserved. During this kind of collision, the energy is transformed into other energy forms like heat and light. Since during the phenomenon the two masses follow the law of conservation of momentum and move in the same direction with same the same speed v we have:

  $$m_1u_1 + m_2u_2 = (m_1 + m_2)v$$

  $$v = (m_1u_1 + m_2u_2)/(m_1 + m_2)$$

  - The kinetic energy of the masses before the collision is : $K.E_1 = 1/2\, m_1u^2_1 + 1/2\, m_2u^2_2$.
  - While kinetic energy after the collision is: $K.E_2 = 1/2\,(m_1 + m_2)\,v^2$.
  - But according to the law of conservation of energy: $1/2\, m_1u^2_1 + 1/2\, m_2u^2_2 = 1/2\,(m_1 + m_2)\,v^2 + Q$.
  - 'Q' here is the change in energy that results in the production of heat or sound.

The Coefficient of Restitution

The coefficient of restitution is the ratio between the relative velocity of colliding masses before interaction to the relative velocity of the masses after the collision. Represented by 'e', the coefficient of restitution depends on the material of the colliding masses. For elastic collisions, e = 1 while for inelastic collisions,e = 0. The value of e > 0 or e < 1 in all other kinds of forceful interactions.

One Dimensional Collision

One dimensional sudden interaction of masses is that collision in which both the initial and final velocities of the masses lie in one line. All the variables of motion are contained in a single dimension.

Elastic One Dimensional Collision

In the elastic collisions the internal kinetic energy is conserved so is the momentum. Elastic collisions can be achieved only with particles like microscopic particles like electrons, protons or neutrons.

$$m_1u_1 + m_2u_2 = m_1v_1 + m_2v_2$$

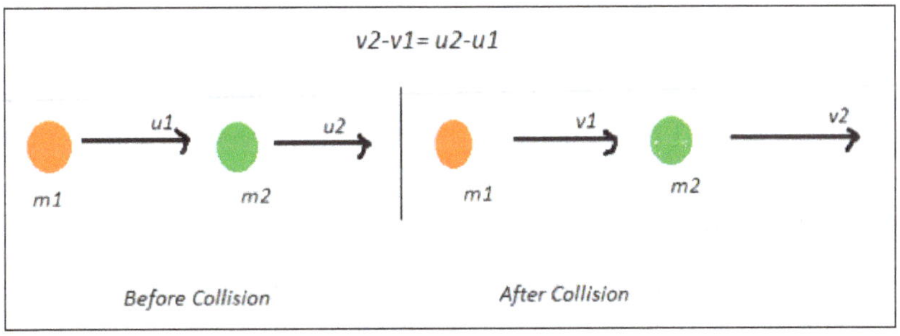

Since the kinetic energy is conserved in the elastic collision we have:

$$1/2\ m_1u^2_1 + 1/2\ m_2u^2_2 = 1/2\ m_1v^2_1 + 1/2\ m_2v^2_2$$

This gives us : $m_1u^2_1 + m_2u^2_2 = m_1v^2_1 + m_2v^2_2$ (Factoring out 1/2).

Rearranging we get: $m_1u^2_1 - m_1v^2_1 = m_2v^2_2 - m_2u^2_2$.

Therefore, $m_1(u^2_1 - v^2_1) = m_2(v^2_2 - u^2_2)$.

Which if elaborated become $m_1(u_1+v_1)(u_1 - v_1) = m_2(v_2 + u_2)(v_2 - u_2)$.

Using the conservation of momentum equation: $m_1u_1 + m_2u_2 = m_1v_1 + m_2v_2$.

We regroup it with same masses: $m_1u_1 - m_1v_1 = m_2v_2 - m_2u_2$.

Hence, $m_1(u_1-v_1) = m_2(v_2-u_2)$.

Now dividing the two equations:

$$m_1(u_1+v_1)(u_1 - v_1) = m_2(v_2 + u_2)(v_2 - u_2) / m_1(u_1-v_1) = m_2(v_2-u_2)$$

We get:

$$u_1+v_1 = v_2 + u_2$$

Now, $v_1 = v_2 + u_2 - u_1$

When we use this value of v_1 in equation of conservation momentum we get:

$$v_2 = [2\, m_1 u_1 + u_2 (m_2 - m_1)] / (m_1 + m_2)$$

Now using the value of v_2 in equation $v_1 = v_2 + u_2 - u_1$,

$$v_1 = [2\, m_1 u_1 + u_2 (m_2 - m_1)] / (m_1 + m_2) + u_2 - u_1$$

$$v_1 = [2\, m_1 u_1 + u_2 (m_2 - m_1) + u_2 (m_1 + m_2) - u_1 (m_1 + m_2)] / (m_1 + m_2)$$

We finally get:

$$v_1 = [2m_2 u_2 + u_1 (m_1 - m_2)] / (m_1 + m_2)$$

When masses of both the bodies are equal then generally after collision, these masses exchange their velocities.

$$v_1 = u_2 \text{ and } v_2 = u_1$$

This means that incourse of collision between objects of same masses, if the second mass is at rest and the firstmass collides with it then after collision the first mass comes to rest and the second mass moves with the speed equal to first mass. Therefore in such case, $v_1 = 0$ and $v_2 = u_1$. In case if $m_1 < m_2$ then, $v_1 = -u_1$ and $v_2 = 0$.

This means that the lighter body will bombard back with its own velocity, while the heavier mass will remain static. However, if $m_1 > m_2$ then $v_1 = u_1$ and $v_2 = 2u_1$.

Inelastic One Dimensional Collision

In inelastic one dimensional collision, the colliding masses stick together and move in the same direction at same speeds. The momentum is conserved and Kinetic energy is changed to different forms of energies. For inelastic collisions the equation for conservation of momentum is :

$$m_1 u_1 + m_2 u_2 = (m_1 + m_2) v$$

Since both the objects stick, we take final velocity after the collision as v. Now v shall be:

$$= m_1 u_1 + m_2 u_2 / m_1 + m_2$$

The kinetic energy lost during the phenomenon shall be:

$$E = 1/2\, m_1 u^2_2 - 1/2\, (m_1 + m_2)\, v^2$$

Collision in Two Dimensions

The below figure signifies collision in two dimensions, where the masses move in different directions after colliding. Here the moving mass m_1 collides with stationary mass m_2. The linear momentum is conserved in the two-dimensional interaction of masses. In this case, we see the masses moving in x,y planes. The x and y component equations are:

$$m_1u_1 = m_1u_2\cos\theta_1 + m_2v_2\cos\theta_2$$

$$0 = m_1u_2\sin\theta_1 - m_2v_2\sin\theta_2.$$

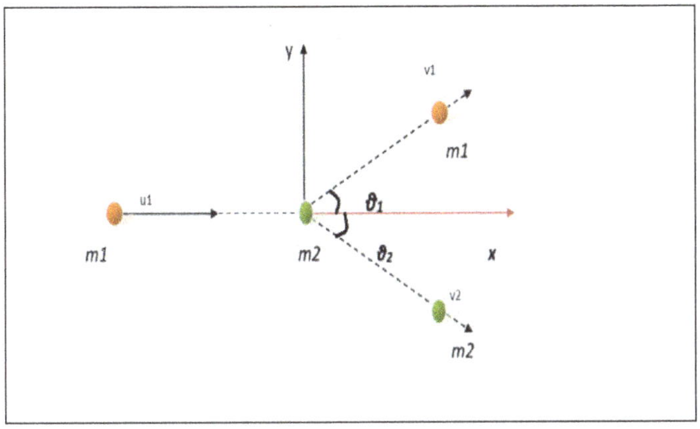

For spherical objects that have smooth surfaces, the collision takes place only when the objects touch with each other. This is what happens in the games of marbles, carom, and billiards.

References

- Mechanical-work, entry: newworldencyclopedia.org, Retrieved 5 May, 2019
- Andrew Pytel; Jaan Kiusalaas (2010). Engineering Mechanics: Dynamics – SI Version, Volume 2 (3rd ed.). Cengage Learning. P. 654. ISBN 9780495295631
- Energy, science:britannica.com, Retrieved 26 July, 2019
- School of Mathematics and Statistics, University of St Andrews (2000). “Biography of Gaspard-Gustave de Coriolis (1792-1843)”. Retrieved 2006-03-03
- Tipler, Paul (2004). Physics for Scientists and Engineers: Mechanics, Oscillations and Waves, Thermodynamics (5th ed.). W. H. Freeman. ISBN 0-7167-0809-4
- Carl Nave (2010). “Elastic and inelastic collisions”. Hyperphysics. Archived from the original on 18 August 2012. Retrieved 2 August 2012
- Collision, science:britannica.com, Retrieved 8 January, 2019
- Collisions, work-energy-and-power, physics, guides: toppr.com, Retrieved 16 January, 2019

6

Gravitation

The universal force of attraction which acts between all matter with mass and energy including planets, stars, galaxies, light, etc. is referred to as gravitation. It includes gravitational constant, gravitational acceleration, gravitational potential, etc. All these diverse concepts of gravitation have been carefully analyzed in this chapter.

In mechanics, gravity (also called gravitation) is the universal force of attraction acting between all matter. It is by far the weakest known force in nature and thus plays no role in determining the internal properties of everyday matter. On the other hand, through its long reach and universal action, it controls the trajectories of bodies in the solar system and elsewhere in the universe and the structures and evolution of stars, galaxies, and the whole cosmos. On Earth all bodies have a weight, or downward force of gravity, proportional to their mass, which Earth's mass exerts on them. Gravity is measured by the acceleration that it gives to freely falling objects. At Earth's surface the acceleration of gravity is about 9.8 metres (32 feet) per second per second. Thus, for every second an object is in free fall, its speed increases by about 9.8 metres per second. At the surface of the Moon the acceleration of a freely falling body is about 1.6 metres per second per second.

The works of Isaac Newton and Albert Einstein dominate the development of gravitational theory. Newton's classical theory of gravitational force held sway from his *Principia*, published in 1687, until Einstein's work in the early 20th century. Newton's theory is sufficient even today for all but the most precise applications. Einstein's theory of general relativity predicts only minute quantitative differences from the Newtonian theory except in a few special cases. The major significance of Einstein's theory is its radical conceptual departure from classical theory and its implications for further growth in physical thought.

The launch of space vehicles and developments of research from them have led to great improvements in measurements of gravity around Earth, other planets, and the Moon and in experiments on the nature of gravitation.

Newton's Law of Universal Gravitation

Newton's law of universal gravitation states that every particle attracts every other particle in the universe with a force which is directly proportional to the product of their masses and inversely proportional to the square of the distance between their centers.This is a general physical law derived from empirical observations by what Isaac Newton called inductive reasoning. It is a part of classical mechanics and was formulated in Newton's work Philosophiæ Naturalis Principia Mathematica ("the Principia"), first published on 5 July 1687.

In today's language, the law states that every point mass attracts every other point mass by a force acting along the line intersecting the two points. The force is proportional to the product of the two masses, and inversely proportional to the square of the distance between them.

The equation for universal gravitation thus takes the form:

$$F = G\frac{m_1 m_2}{r^2},$$

where F is the gravitational force acting between two objects, m_1 and m_2 are the masses of the objects, r is the distance between the centers of their masses, and G is the gravitational constant.

The first test of Newton's theory of gravitation between masses in the laboratory was the Cavendish experiment conducted by the British scientist Henry Cavendish in 1798. It took place 111 years after the publication of Newton's Principia and approximately 71 years after his death.

Newton's law of gravitation resembles Coulomb's law of electrical forces, which is used to calculate the magnitude of the electrical force arising between two charged bodies. Both are inverse-square laws, where force is inversely proportional to the square of the distance between the bodies. Coulomb's law has the product of two charges in place of the product of the masses, and the electrostatic constant in place of the gravitational constant.

Newton's law has since been superseded by Albert Einstein's theory of general relativity, but it continues to be used as an excellent approximation of the effects of gravity in most applications. Relativity is required only when there is a need for extreme accuracy, or when dealing with very strong gravitational fields, such as those found near extremely massive and dense objects, or at very close distances (such as Mercury's orbit around the Sun).

Modern Form

In modern language, the law states the following:

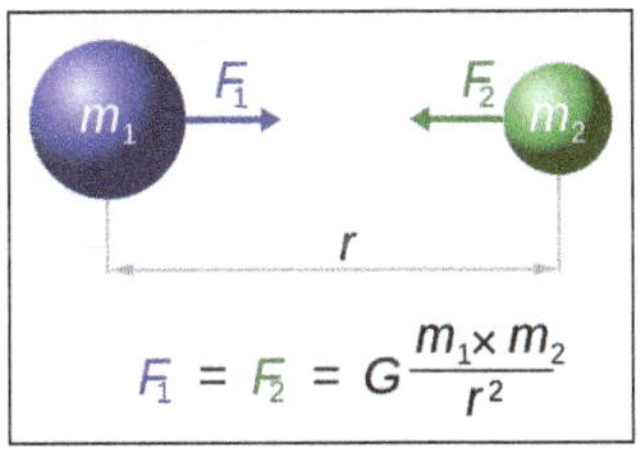

Every point mass attracts every single other point mass by a force acting along the line intersecting both points. The force is proportional to the product of the two masses and inversely proportional to the square of the distance between them,

$$F = G\frac{m_1 m_2}{r^2}$$

where:

F is the force between the masses;

G is the gravitational constant (6.674×10^{-11} N·(m/kg)²);

m_1 is the first mass;

m_2 is the second mass;

r is the distance between the centers of the masses.

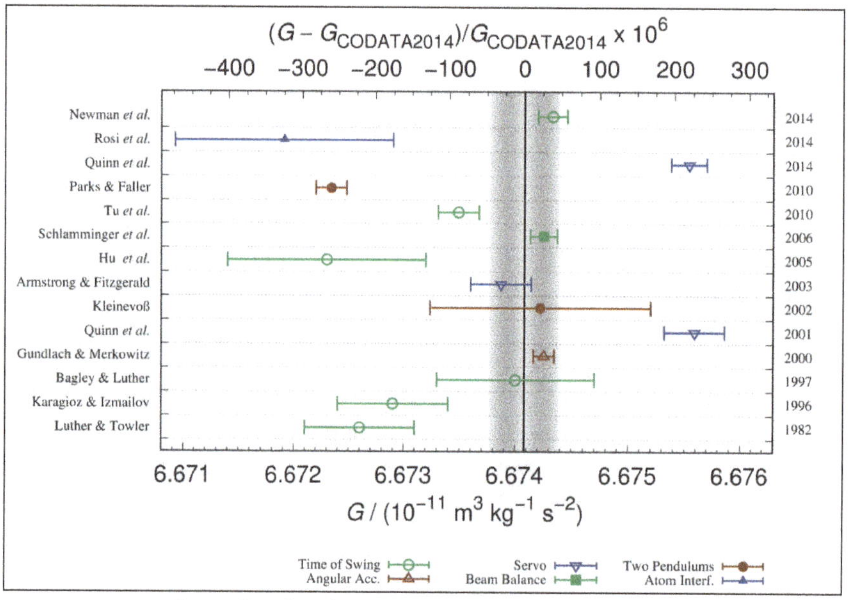

Error plot showing experimental values for big G.

Assuming SI units, F is measured in newtons (N), m_1 and m_2 in kilograms (kg), r in meters (m), and the constant G is approximately equal to 6.674×10^{-11} N m² kg⁻². The value of the constant G was first accurately determined from the results of the Cavendish experiment conducted by the British scientist Henry Cavendish in 1798, although Cavendish did not himself calculate a numerical value for G. This experiment was also the first test of Newton's theory of gravitation between masses in the laboratory. It took place 111 years after the publication of Newton's *Principia* and 71 years after Newton's death, so none of Newton's calculations could use the value of G; instead he could only calculate a force relative to another force.

Bodies with Spatial Extent

If the bodies in question have spatial extent (as opposed to being point masses), then the gravitational force between them is calculated by summing the contributions of the notional point masses which constitute the bodies. In the limit, as the component point masses become "infinitely small", this entails integrating the force over the extents of the two bodies.

In this way, it can be shown that an object with a spherically-symmetric distribution of mass exerts the same gravitational attraction on external bodies as if all the object's mass were concentrated at a point at its center. (This is not generally true for non-spherically-symmetrical bodies.)

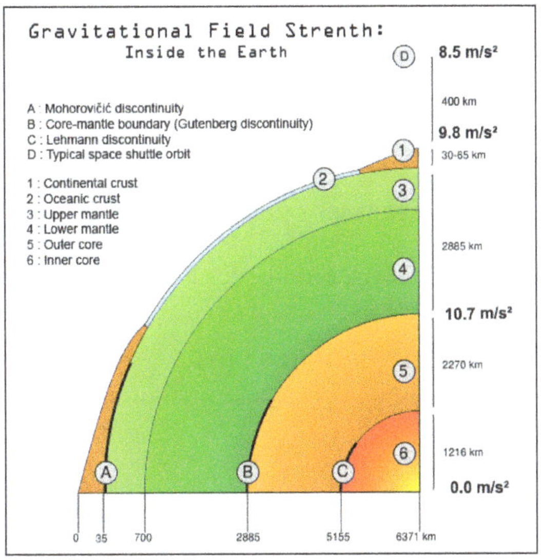

Gravitational field strength within the Earth.

For points *inside* a spherically-symmetric distribution of matter, Newton's Shell theorem can be used to find the gravitational force. The theorem tells us how different parts of the mass distribution affect the gravitational force measured at a point located a distance $\mathbf{r}_0$ from the center of the mass distribution:

- The portion of the mass that is located at radii $r < \mathbf{r}_0$ causes the same force at $\mathbf{r}_0$ as if all of the mass enclosed within a sphere of radius $\mathbf{r}_0$ was concentrated at the center of the mass distribution (as noted above).
- The portion of the mass that is located at radii $r > \mathbf{r}_0$ exerts *no net* gravitational force at the distance $\mathbf{r}_0$ from the center. That is, the individual gravitational forces exerted by the elements of the sphere out there, on the point at $\mathbf{r}_0$, cancel each other out.

As a consequence, for example, within a shell of uniform thickness and density there is *no net* gravitational acceleration anywhere within the hollow sphere.

Furthermore, inside a uniform sphere the gravity increases linearly with the distance from the center; the increase due to the additional mass is 1.5 times the decrease due to the larger distance from the center. Thus, if a spherically symmetric body has a uniform core and a uniform mantle with a density that is less than 2/3 of that of the core, then the gravity initially decreases outwardly beyond the boundary, and if the sphere is large enough, further outward the gravity increases again, and eventually it exceeds the gravity at the core/mantle boundary. The gravity of the Earth may be highest at the core/mantle boundary.

Vector Form

Newton's law of universal gravitation can be written as a vector equation to account for the direction of the gravitational force as well as its magnitude. In this formula, quantities in bold represent vectors.

$$\mathbf{F}_{21} = -G\frac{m_1 m_2}{|\mathbf{r}_{12}|^2}\hat{\mathbf{r}}_{12}$$

where

> $\mathbf{F}_{21}$ is the force applied on object 2 exerted by object 1,
>
> G is the gravitational constant,
>
> m_1 and m_2 are respectively the masses of objects 1 and 2,
>
> $|\mathbf{r}_{12}| = |\mathbf{r}_2 - \mathbf{r}_1|$ is the distance between objects 1 and 2, and
>
> $\hat{\mathbf{r}}_{12} \stackrel{\text{def}}{=} \frac{\mathbf{r}_2 - \mathbf{r}_1}{|\mathbf{r}_2 - \mathbf{r}_1|}$ is the unit vector from object 1 to 2.

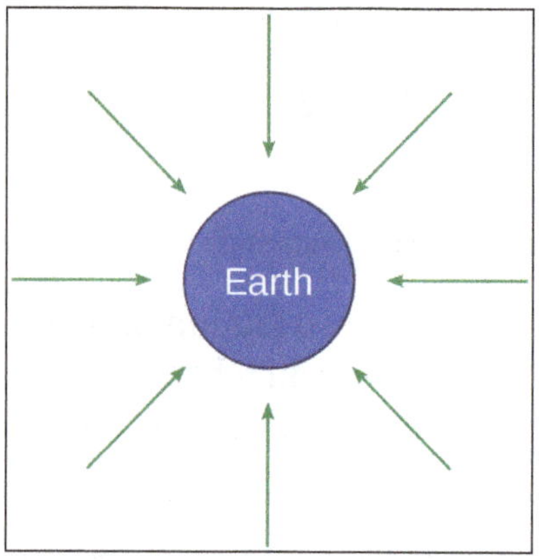

Gravity field surrounding Earth from a macroscopic perspective.

It can be seen that the vector form of the equation is the same as the scalar form given earlier, except that **F** is now a vector quantity, and the right hand side is multiplied by the appropriate unit vector. Also, it can be seen that $\mathbf{F}_{12} = -\mathbf{F}_{21}$.

Gravitational Field

The gravitational field is a vector field that describes the gravitational force which would be applied on an object in any given point in space, per unit mass. It is actually equal to the gravitational acceleration at that point.

It is a generalisation of the vector form, which becomes particularly useful if more than 2 objects are involved (such as a rocket between the Earth and the Moon). For 2 objects (e.g. object 2 is a rocket, object 1 the Earth), we simply write **r** instead of $\mathbf{r}_{12}$ and m instead of m_2 and define the gravitational field **g(r)** as:

$$\mathbf{g}(\mathbf{r}) = -G\frac{m_1}{|\mathbf{r}|^2}\hat{\mathbf{r}}$$

so that we can write:

$$\mathbf{F}(\mathbf{r}) = m\mathbf{g}(\mathbf{r}).$$

This formulation is dependent on the objects causing the field. The field has units of acceleration; in SI, this is m/s^2.

Gravitational fields are also conservative; that is, the work done by gravity from one position to another is path-independent. This has the consequence that there exists a gravitational potential field $V(\mathbf{r})$ such that,

$$\mathbf{g}(\mathbf{r}) = -\nabla V(\mathbf{r}).$$

If m_1 is a point mass or the mass of a sphere with homogeneous mass distribution, the force field $\mathbf{g}(\mathbf{r})$ outside the sphere is isotropic, i.e., depends only on the distance r from the center of the sphere. In that case,

$$V(r) = -G\frac{m_1}{r}.$$

the gravitational field is on, inside and outside of symmetric masses.

As per Gauss's law, field in a symmetric body can be found by the mathematical equation:

$$\oiint_{\partial V} \mathbf{g}(\mathbf{r}) \cdot d\mathbf{A} = -4\pi G M_{\text{enc}},$$

where ∂V is a closed surface and M_{enc} is the mass enclosed by the surface.

Hence, for a hollow sphere of radius R and total mass M,

$$|\mathbf{g}(\mathbf{r})| = \begin{cases} 0, & \text{if } r < R \\ \dfrac{GM}{r^2}, & \text{if } r \geq R \end{cases}$$

For a uniform solid sphere of radius R and total mass M,

$$|\mathbf{g}(\mathbf{r})| = \begin{cases} \dfrac{GMr}{R^3}, & \text{if } r < R \\ \dfrac{GM}{r^2}, & \text{if } r \geq R \end{cases}$$

Problematic Aspects

Newton's description of gravity is sufficiently accurate for many practical purposes and is therefore widely used. Deviations from it are small when the dimensionless quantities ϕ / c^2 and $(v / c)^2$ are both much less than one, where ϕ is the gravitational potential, v is the velocity of the objects being studied, and c is the speed of light in vacuum. For example, Newtonian gravity provides an accurate description of the Earth/Sun system, since

$$\frac{\phi}{c^2} = \frac{GM_{\text{sun}}}{r_{\text{orbit}} c^2} \sim 10^{-8}, \quad \left(\frac{v_{\text{Earth}}}{c}\right)^2 = \left(\frac{2\pi r_{\text{orbit}}}{(1\text{ yr})c}\right)^2 \sim 10^{-8}$$

where r_{orbit} is the radius of the Earth's orbit around the Sun.

In situations where either dimensionless parameter is large, then general relativity must be used to describe the system. General relativity reduces to Newtonian gravity in the limit of small potential and low velocities, so Newton's law of gravitation is often said to be the low-gravity limit of general relativity.

Theoretical Concerns with Newton's Expression

- There is no immediate prospect of identifying the mediator of gravity. Attempts by physicists to identify the relationship between the gravitational force and other known fundamental forces are not yet resolved, although considerable headway has been made over the last 50 years. Newton himself felt that the concept of an inexplicable *action at a distance* was unsatisfactory, but that there was nothing more that he could do at the time.
- Newton's theory of gravitation requires that the gravitational force be transmitted instantaneously. Given the classical assumptions of the nature of space and time before the development of General Relativity, a significant propagation delay in gravity leads to unstable planetary and stellar orbits.

Observations Conflicting with Newton's Formula

- Newton's theory does not fully explain the precession of the perihelion of the orbits of the planets, especially of planet Mercury, which was detected long after the life of Newton. There is a 43 arcsecond per century discrepancy between the Newtonian calculation, which arises only from the gravitational attractions from the other planets, and the observed precession, made with advanced telescopes during the 19th century.
- The predicted angular deflection of light rays by gravity that is calculated by using Newton's Theory is only one-half of the deflection that is actually observed by astronomers. Calculations using general relativity are in much closer agreement with the astronomical observations.
- In spiral galaxies, the orbiting of stars around their centers seems to strongly disobey Newton's law of universal gravitation. Astrophysicists, however, explain this spectacular phenomenon in the framework of Newton's laws, with the presence of large amounts of dark matter.

Newton's Reservations

While Newton was able to formulate his law of gravity in his monumental work, he was deeply uncomfortable with the notion of "action at a distance" that his equations implied. In 1692, in his third letter to Bentley, he wrote: "That one body may act upon another at a distance through a vacuum without the mediation of anything else, by and through which their action and force may be conveyed from one another, is to me so great an absurdity that, I believe, no man who has in philosophic matters a competent faculty of thinking could ever fall into it."

He never, in his words, "assigned the cause of this power". In all other cases, he used the phenomenon of motion to explain the origin of various forces acting on bodies, but in the case of gravity, he was unable to experimentally identify the motion that produces the force of gravity (although he invented two mechanical hypotheses in 1675 and 1717). Moreover, he refused to even offer a hypothesis as to the cause of this force on grounds that to do so was contrary to sound science. He lamented that "philosophers have hitherto attempted the search of nature in vain" for the source of the gravitational force, as he was convinced "by many reasons" that there were "causes hitherto unknown" that were fundamental to all the "phenomena of nature". These fundamental phenomena are still under investigation and, though hypotheses abound, the definitive answer has yet to be found. And in Newton's 1713 General Scholium in the second edition of Principia: "I have not yet been able to discover the cause of these properties of gravity from phenomena and I feign no hypotheses. It is enough that gravity does really exist and acts according to the laws I have explained, and that it abundantly serves to account for all the motions of celestial bodies."

Einstein's Solution

These objections were explained by Einstein's theory of general relativity, in which gravitation is an attribute of curved spacetime instead of being due to a force propagated between bodies. In Einstein's theory, energy and momentum distort spacetime in their vicinity, and other particles move in trajectories determined by the geometry of spacetime. This allowed a description of the motions of light and mass that was consistent with all available observations. In general relativity, the gravitational force is a fictitious force due to the curvature of spacetime, because the gravitational acceleration of a body in free fall is due to its world line being a geodesic of spacetime.

Extensions

Newton was the first to consider in his *Principia* an extended expression of his law of gravity including an inverse-cube term of the form,

$$F = G\frac{m_1 m_2}{r^2} + B\frac{m_1 m_2}{r^3}, \qquad B \text{ a constant}$$

attempting to explain the Moon's apsidal motion. Other extensions were proposed by Laplace (around 1790) and Decombes (1913):

$$F(r) = k\frac{m_1 m_2}{r^2}\exp(-\alpha r) \qquad \text{(Laplace)}$$

$$F(r) = k\frac{m_1 m_2}{r^2}\left(1 + \frac{\alpha}{r^3}\right) \qquad \text{(Decombes)}$$

In recent years, quests for non-inverse square terms in the law of gravity have been carried out by neutron interferometry.

Solutions of Newton's Law of Universal Gravitation

The *n*-body problem is an ancient, classical problem of predicting the individual motions of a group of celestial objects interacting with each other gravitationally. Solving this problem — from the time of the Greeks and on — has been motivated by the desire to understand the motions of the Sun, planets and the visible stars. In the 20th century, understanding the dynamics of globular cluster star systems became an important n-body problem too. The n-body problem in general relativity is considerably more difficult to solve.

The classical physical problem can be informally stated as: given the quasi-steady orbital properties (instantaneous position, velocity and time) of a group of celestial bodies, predict their interactive forces; and consequently, predict their true orbital motions for all future times.

The two-body problem has been completely solved, as has the restricted three-body problem.

Gravitational Constant

Values of G	Units
$6.67430(15)\times10^{-11}$	$m^3 \cdot kg^{-1} \cdot s^{-2}$
$4.30091(25)\times10^{-3}$	$pc \cdot M_{\odot}^{-1} \cdot (km/s)^2$

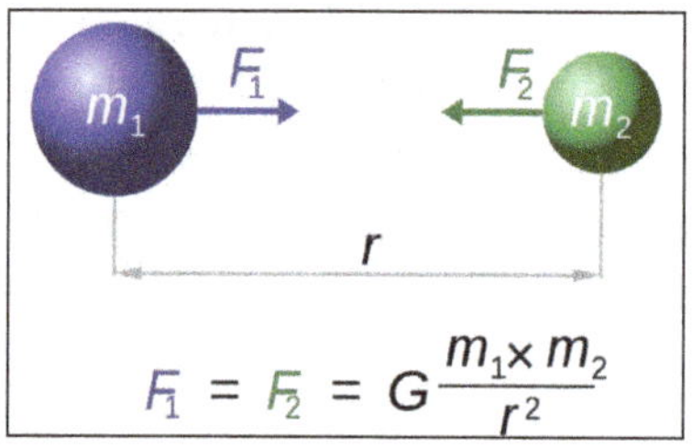

The gravitational constant G is a key quantity in Newton's law of universal gravitation.

The gravitational constant (also known as the universal gravitational constant, the Newtonian constant of gravitation, or the Cavendish gravitational constant), denoted by the letter G, is an empirical physical constant involved in the calculation of gravitational effects in Sir Isaac Newton's law of universal gravitation and in Albert Einstein's general theory of relativity.

In Newton's law, it is the proportionality constant connecting the gravitational force between two bodies with the product of their masses and the inverse square of their distance. In the Einstein field equations, it quantifies the relation between the geometry of spacetime and the energy–momentum tensor.

The measured value of the constant is known with some certainty to four significant digits. In SI units its value is approximately 6.674×10^{-11} $m^3 \cdot kg^{-1} \cdot s^{-2}$.

The modern notation of Newton's law involving G was introduced in the 1890s by C. V. Boys. The first implicit measurement with an accuracy within about 1% is attributed to Henry Cavendish in a 1798 experiment.

According to Newton's law of universal gravitation, the attractive force (F) between two point-like bodies is directly proportional to the product of their masses (m_1 and m_2), and inversely proportional to the square of the distance, r, (inverse-square law) between them:

$$F = G\frac{m_1 m_2}{r^2}.$$

The constant of proportionality, G, is the gravitational constant. Colloquially, the gravitational constant is also called "Big G", for disambiguation with "small g" (g), which is the local gravitational field of Earth (equivalent to the free-fall acceleration). Where $M_\oplus$ is the mass of the Earth and $r_\oplus$ is the radius of the Earth, the two quantities are related by:

$$g = \frac{GM_\oplus}{r_\oplus^2}$$

In the Einstein field equations of general relativity,

$$R_{\mu\nu} - \tfrac{1}{2}Rg_{\mu\nu} = \frac{8\pi G}{c^4}T_{\mu\nu},$$

Newton's constant appears in the proportionality between the spacetime curvature and the energy density component of the stress–energy tensor. The scaled gravitational constant, or Einstein's constant, is:

$$\kappa = \frac{8\pi}{c^4}G \approx 2.071\times 10^{-43}\ \mathrm{s^2\cdot m^{-1}\cdot kg^{-1}}.$$

Value and Dimensions

The gravitational constant is a physical constant that is difficult to measure with high accuracy. This is because the gravitational force is extremely weak as compared to other fundamental forces.

In SI units, the 2018 CODATA-recommended value of the gravitational constant (with standard uncertainty in parentheses) is:

$$G = 6.67430(15)\times 10^{-11}\ \mathrm{m^3{\cdot}kg^{-1}{\cdot}s^{-2}}$$

This corresponds to a relative standard uncertainty of 2.2×10^{-5} (22 ppm).

The dimensions assigned to the gravitational constant are force times length squared divided by mass squared; this is equivalent to length cubed, divided by mass and by time squared:

$$[G] = \frac{[F][L]^2}{[M]^2} = \frac{[L]^3}{[M][T]^2}$$

In SI base units, this amounts to meters cubed per kilogram per second squared:

$$\mathrm{N{\cdot}m^2{\cdot}kg^{-2}} = \mathrm{m^3{\cdot}kg^{-1}{\cdot}s^{-2}}.$$

In cgs, G can be written as $G \approx 6.674\times10^{-8}\ \mathrm{cm^3{\cdot}g^{-1}{\cdot}s^{-2}}$.

Natural Units

The gravitational constant is taken as the basis of the Planck units: it is equal to the cube of the Planck length divided by the product of the Planck mass and the square of Planck time:

$$G = \frac{l_P^3}{m_P t_P^2}.$$

In other words, in Planck units, G has the numerical value of 1.

Thus, in Planck units, and other natural units taking G as their basis, the value of the gravitational constant cannot be measured as this is set by definition. Depending on the choice of units, uncertainty in the value of a physical constant as expressed in one system of units shows up as uncertainty of the value of another constant in another system of units. Where there is variation in dimensionless physical constants, no matter which choice of physical "constants" is used to define the units, this variation is preserved independently of the choice of units; in the case of the gravitational constant, such a dimensionless value is the gravitational coupling constant of the electron,

$$\alpha_G = \frac{Gm_e^2}{\hbar c} = \left(\frac{m_e}{m_P}\right)^2 \approx 1.7518 \times 10^{-45},$$

a measure for the gravitational attraction between a pair of electrons, proportional to the square of the electron rest mass.

Orbital Mechanics

In astrophysics, it is convenient to measure distances in parsecs (pc), velocities in kilometers per second (km/s) and masses in solar units $M_\odot$. In these units, the gravitational constant is:

$$G \approx 4.302 \times 10^{-3} \text{ pc } M_\odot^{-1} \text{ (km/s)}^2.$$

For situations where tides are important, the relevant length scales are solar radii rather than parsecs. In these units, the gravitational constant is:

$$G \approx 1.90809 \times 10^5 R_\odot M_\odot^{-1} \text{ (km/s)}^2.$$

In orbital mechanics, the period P of an object in circular orbit around a spherical object obeys,

$$GM = \frac{3\pi V}{P^2}$$

where V is the volume inside the radius of the orbit. It follows that,

$$P^2 = \frac{3\pi}{G}\frac{V}{M} \approx 10.896 \text{ hr}^2 \cdot \text{g} \cdot \text{cm}^{-3} \frac{V}{M}.$$

This way of expressing G shows the relationship between the average density of a planet and the period of a satellite orbiting just above its surface.

For elliptical orbits, applying Kepler's 3rd law, expressed in units characteristic of Earth's orbit:

$$G = 4\pi^2 \text{ AU}^3 \cdot \text{yr}^{-2} M^{-1} \approx 39.478 \text{ AU}^3 \cdot \text{yr}^{-2} M_\odot^{-1},$$

where distance is measured in terms of the semi-major axis of Earth's orbit (the astronomical unit, AU), time in years, and mass in the total mass of the orbiting system ($M = M_\odot + M_\oplus + M_☾$).

The above equation is exact only within the approximation of the Earth's orbit around the Sun as a two-body problem in Newtonian mechanics, the measured quantities contain corrections from the perturbations from other bodies in the solar system and from general relativity.

From 1964 until 2012, however, it was used as the definition of the astronomical unit and thus held by definition:

$$1\,\mathrm{AU} = \left(\frac{GM}{\quad}\,\mathrm{yr}^2\right)^{-} \approx 1.495979\times10^{11}\,\mathrm{m}.$$

Since 2012, the AU is defined as 1.495978707×10^{11} m exactly, and the equation can no longer be taken as holding precisely.

The quantity *GM*—the product of the gravitational constant and the mass of a given astronomical body such as the Sun or Earth—is known as the standard gravitational parameter and (also denoted μ). The standard gravitational parameter *GM* appears as above in Newton's law of universal gravitation, as well as in formulas for the deflection of light caused by gravitational lensing, in Kepler's laws of planetary motion, and in the formula for escape velocity.

This quantity gives a convenient simplification of various gravity-related formulas. The product *GM* is known much more accurately than either factor is.

Values for GM			
Body	μ = GM	Value	Precision
Sun	$GM_\odot$	$1.32712440018(9)\times10^{20}\ \mathrm{m^3{\cdot}s^{-2}}$	10 digits
Earth	$GM_\oplus$	$3.986004418(8)\times10^{14}\ \mathrm{m^3{\cdot}s^{-2}}$	9 digits

Calculations in celestial mechanics can also be carried out using the units of solar masses, mean solar days and astronomical units rather than standard SI units. For this purpose, the Gaussian gravitational constant was historically in widespread use, $k = 0.01720209895$, expressing the mean angular velocity of the Sun–Earth system measured in radians per day.

Gravitational Acceleration

In physics, gravitational acceleration is the acceleration on an object caused by the force of gravitation. Neglecting friction such as air resistance, all small bodies accelerate in a gravitational field at the same rate relative to the center of mass. This equality is true regardless of the masses or compositions of the bodies.

At different points on Earth, objects fall with an acceleration between 9.764 m/s^2 and 9.834 m/s^2 depending on altitude and latitude, with a conventional standard value of exactly 9.80665 m/s^2 (approximately 32.17405 ft/s^2). This does not take into account other effects, such as buoyancy or drag.

For Point Masses

Newton's law of universal gravitation states that there is a gravitational force between any two masses that is equal in magnitude for each mass, and is aligned to draw the two masses toward each other. The formula is:

$$F = G\frac{m_1 m_2}{r^2}$$

where m_1 and m_2 are the two masses, G is the gravitational constant, and r is the distance between the two masses. The formula was derived for planetary motion where the distances between the planets and the Sun made it reasonable to consider the bodies to be point masses. (For a satellite in orbit, the 'distance' refers to the distance from the mass centers rather than, say, the altitude above a planet's surface.)

If one of the masses is much larger than the other, it is convenient to define a gravitational field around the larger mass as follows:

$$\mathbf{g} = -\frac{GM}{r^2}\hat{\mathbf{r}}$$

where M is the mass of the larger body, and $\hat{\mathbf{r}}$ is a unit vector directed from the large mass to the smaller mass. The negative sign indicates that the force is an attractive force.

In that way, the force acting upon the smaller mass can be calculated as:

$$\mathbf{F} = m\mathbf{g}$$

where **F** is the force vector, m is the smaller mass, and **g** is a vector pointing toward the larger body. **g** has units of acceleration and is a vector function of location relative to the large body, independent of the magnitude (or even the presence) of the smaller mass.

This model represents the "far-field" gravitational acceleration associated with a massive body. When the dimensions of a body are not trivial compared to the distances of interest, the principle of superposition can be used for differential masses for an assumed density distribution throughout the body in order to get a more detailed model of the "near-field" gravitational acceleration. For satellites in orbit, the far-field model is sufficient for rough calculations of altitude versus period, but not for precision estimation of future location after multiple orbits.

The more detailed models include (among other things) the bulging at the equator for the Earth, and irregular mass concentrations (due to meteor impacts) for the Moon. The Gravity Recovery And Climate Experiment (GRACE) mission launched in 2002 consists of two probes, nicknamed "Tom" and "Jerry", in polar orbit around the Earth measuring differences in the distance between the two probes in order to more precisely determine the gravitational field around the Earth, and to track changes that occur over time. Similarly, the Gravity Recovery and Interior Laboratory (GRAIL) mission from 2011-2012 consisted of two probes ("Ebb" and "Flow") in polar orbit around the Moon to more precisely determine the gravitational field for future navigational purposes, and to infer information about the Moon's physical makeup.

Gravity Model for Earth

The type of gravity model used for the Earth depends upon the degree of fidelity required for a given problem. For many problems such as aircraft simulation, it may be sufficient to consider gravity to be a constant, defined as:

$$g = 9.80665 \text{ metres (32.1740 ft) per s}^2$$

where is understood to be pointing 'down' in the local frame of reference.

If it is desirable to model an object's weight on Earth as a function of latitude, one could use the following:

$$g = g_{45} - \tfrac{1}{2}(g_{\text{poles}} - g_{\text{equator}})\cos\left(2\varphi\cdot\frac{\pi}{180}\right)$$

where

$g_{\text{poles}} = 9.832$ metres (32.26 ft) per s^2,

$g_{45} = 9.806$ metres (32.17 ft) per s^2,

$g_{\text{equator}} = 9.780$ metres (32.09 ft) per s^2,

$\varphi =$ latitude, between −90 and 90 degrees.

Neither of these accounts for changes in gravity with changes in altitude, but the model with the cosine function does take into account the centrifugal relief that is produced by the rotation of the Earth. For the mass attraction effect by itself, the gravitational acceleration at the equator is about 0.18% less than that at the poles due to being located farther from the mass center. When the rotational component is included (as above), the gravity at the equator is about 0.53% less than that at the poles, with gravity at the poles being unaffected by the rotation. So the rotational component of change due to latitude (0.35%) is about twice as significant as the mass attraction change due to latitude (0.18%), but both reduce strength of gravity at the equator as compared to gravity at the poles.

Note that for satellites, orbits are decoupled from the rotation of the Earth so the orbital period is not necessarily one day, but also that errors can accumulate over multiple orbits so that accuracy is important. For such problems, the rotation of the Earth would be immaterial unless variations with longitude are modeled. Also, the variation in gravity with altitude becomes important, especially for highly elliptical orbits.

The Earth Gravitational Model 1996 (EGM96) contains 130,676 coefficients that refine the model of the Earth's gravitational field. The most significant correction term is about two orders of magnitude more significant than the next largest term. That coefficient is referred to as the J_2 term, and accounts for the flattening of the poles, or the oblateness, of the Earth. (A shape elongated on its axis of symmetry, like an American football, would be called prolate.) A gravitational potential function can be written for the change in potential energy for a unit mass that is brought from infinity into proximity to the Earth. Taking partial derivatives of that function with respect to a coordinate system will then resolve the directional components of the gravitational acceleration vector, as a function of location. The component due to the Earth's rotation can then be included, if appropriate, based on a sidereal day relative to the stars (≈366.24 days/year) rather than on a solar day (≈365.24 days/year). That component is perpendicular to the axis of rotation rather than to the surface of the Earth.

A similar model adjusted for the geometry and gravitational field for Mars can be found in publication NASA SP-8010.

The barycentric gravitational acceleration at a point in space is given by:

$$\mathbf{g} = -\frac{GM}{r^2}\hat{\mathbf{r}}$$

where:

M is the mass of the attracting object, $\hat{\mathbf{r}}$ is the unit vector from center-of-mass of the attracting object to the center-of-mass of the object being accelerated, *r* is the distance between the two objects, and *G* is the gravitational constant.

When this calculation is done for objects on the surface of the Earth, or aircraft that rotate with the Earth, one has to account for the fact that the Earth is rotating and the centrifugal acceleration has to be subtracted from this. For example, the equation above gives the acceleration at 9.820 m/s², when $GM = 3.986\times10^{14}$ m³/s², and $R=6.371\times10^6$ m. The centripetal radius is $r = R\cos(\varphi)$, and the centripetal time unit is approximately $(day / 2\pi)$, reduces this, for $r = 5\times10^6$ metres, to 9.79379 m/s², which is closer to the observed value.

Gravitational Potential

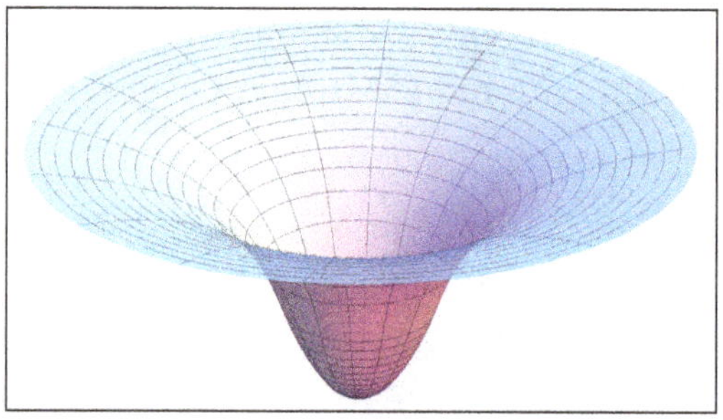

Plot of a two-dimensional slice of the gravitational potential in and around a uniform spherical body. The inflection points of the cross-section are at the surface of the body.

In classical mechanics, the gravitational potential at a location is equal to the work (energy transferred) per unit mass that would be needed to move the object from a fixed reference location to the location of the object. It is analogous to the electric potential with mass playing the role of charge. The reference location, where the potential is zero, is by convention infinitely far away from any mass, resulting in a negative potential at any finite distance.

In mathematics, the gravitational potential is also known as the Newtonian potential and is fundamental in the study of potential theory. It may also be used for solving the electrostatic and magnetostatic fields generated by uniformly charged or polarized ellipsoidal bodies.

Potential Energy

The gravitational potential (V) at a location is the gravitational potential energy (U) at that location per unit mass:

$$V = \frac{U}{m},$$

where m is the mass of the object. Potential energy is equal (in magnitude, but negative) to the work done by the gravitational field moving a body to its given position in space from infinity. If the body has a mass of 1 unit, then the potential energy to be assigned to that body is equal to the gravitational potential. So the potential can be interpreted as the negative of the work done by the gravitational field moving a unit mass in from infinity.

In some situations, the equations can be simplified by assuming a field that is nearly independent of position. For instance, in a region close to the surface of the Earth, the gravitational acceleration, g, can be considered constant. In that case, the difference in potential energy from one height to another is, to a good approximation, linearly related to the difference in height:

$$\Delta U \approx mg\Delta h.$$

Mathematical form

The potential V of a unit mass m at a distance x from a point mass of mass M can be defined as the work W that needs to be done by an external agent to bring the unit mass in from infinity to that point:

$$V(\mathbf{x}) = \frac{W}{m} = \frac{1}{m}\int_{\infty}^{x} \mathbf{F} \cdot d\mathbf{x} = \frac{1}{m}\int_{\infty}^{x} \frac{GmM}{x^2} dx = -\frac{GM}{x},$$

where G is the gravitational constant, and $\mathbf{F}$ is the gravitational force. The potential has units of energy per unit mass, e.g., J/kg in the MKS system. By convention, it is always negative where it is defined, and as x tends to infinity, it approaches zero.

The gravitational field, and thus the acceleration of a small body in the space around the massive object, is the negative gradient of the gravitational potential. Thus the negative of a negative gradient yields positive acceleration toward a massive object. Because the potential has no angular components, its gradient is,

$$\mathbf{a} = -\frac{GM}{x^3}\mathbf{x} = -\frac{GM}{x^2}\hat{\mathbf{x}},$$

where $\mathbf{x}$ is a vector of length x pointing from the point mass toward the small body and $\hat{\mathbf{x}}$ is a unit vector pointing from the point mass toward the small body. The magnitude of the acceleration therefore follows an inverse square law:

$$|\mathbf{a}| = \frac{GM}{x^2}.$$

The potential associated with a mass distribution is the superposition of the potentials of point masses. If the mass distribution is a finite collection of point masses, and if the point masses are located at the points $\mathbf{x}_1, ..., \mathbf{x}_n$ and have masses $m_1, ..., m_n$, then the potential of the distribution at the point $\mathbf{x}$ is,

$$V(\mathbf{x}) = \sum_{i=1}^{n} -\frac{Gm_i}{|\mathbf{x} - \mathbf{x}_i|}.$$

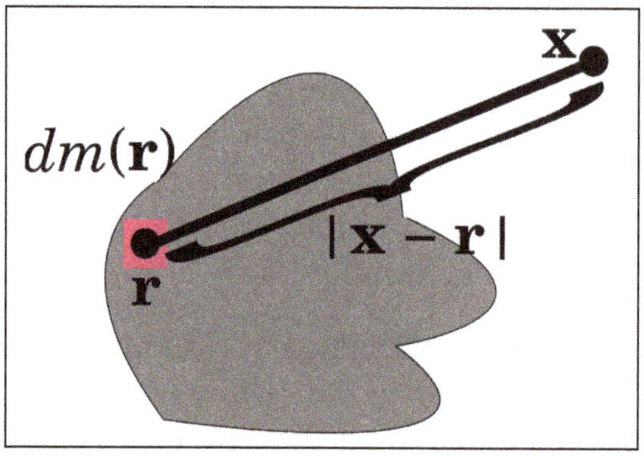

Points **x** and **r**, with **r** contained in the distributed mass (gray) and differential mass *dm*(**r**) located at the point **r**.

If the mass distribution is given as a mass measure *dm* on three-dimensional Euclidean space $\mathbf{R}^3$, then the potential is the convolution of $-G/|\mathbf{r}|$ with *dm*. In good cases this equals the integral

$$V(\mathbf{x}) = -\int_{\mathbf{R}^3} \frac{G}{|\mathbf{x} - \mathbf{r}|} dm(\mathbf{r}),$$

where $|\mathbf{x} - \mathbf{r}|$ is the distance between the points **x** and **r**. If there is a function $\rho(\mathbf{r})$ representing the density of the distribution at **r**, so that $dm(\mathbf{r}) = \rho(\mathbf{r})dv(\mathbf{r})$, where $dv(\mathbf{r})$ is the Euclidean volume element, then the gravitational potential is the volume integral,

$$V(\mathbf{x}) = -\int_{\mathbf{R}^3} \frac{G}{|\mathbf{x} - \mathbf{r}|} \rho(\mathbf{r}) dv(\mathbf{r}).$$

If *V* is a potential function coming from a continuous mass distribution $\rho(\mathbf{r})$, then ρ can be recovered using the Laplace operator, Δ:

$$\rho(\mathbf{x}) = \frac{1}{4\pi G} \Delta V(\mathbf{x}).$$

This holds pointwise whenever ρ is continuous and is zero outside of a bounded set. In general, the mass measure *dm* can be recovered in the same way if the Laplace operator is taken in the sense of distributions. As a consequence, the gravitational potential satisfies Poisson's equation.

The integral may be expressed in terms of known transcendental functions for all ellipsoidal shapes, including the symmetrical and degenerate ones. These include the sphere, where the three semiaxes are equal; the oblate and prolate spheroids, where two semiaxes are equal; the degenerate ones where one semiaxis is infinite (the elliptical and circular cylinder) and the unbounded sheet where two semiaxes are infinite. All these shapes are widely used in the applications of the gravitational potential integral (apart from the constant *G*, with ρ being a constant charge density) to electromagnetism.

Spherical Symmetry

A spherically symmetric mass distribution behaves to an observer completely outside the distribution as though all of the mass was concentrated at the center, and thus effectively as a point mass, by the shell theorem. On the surface of the earth, the acceleration is given by so-called standard gravity g, approximately 9.8 m/s^2, although this value varies slightly with latitude and altitude: The magnitude of the acceleration is a little larger at the poles than at the equator because Earth is an oblate spheroid.

Within a spherically symmetric mass distribution, it is possible to solve Poisson's equation in spherical coordinates. Within a uniform spherical body of radius R and density ρ, the gravitational force g inside the sphere varies linearly with distance r from the center, giving the gravitational potential inside the sphere, which is

$$V(r) = \frac{2}{3}\pi G\rho(r^2 - 3R^2), \qquad r \le R,$$

which differentiably connects to the potential function for the outside of the sphere.

Multipole Expansion

The potential at a point **x** is given by,

$$V(\mathbf{x}) = -\int_{\mathbb{R}^3} \frac{G}{|\mathbf{x}-\mathbf{r}|}\, dm(\mathbf{r}).$$

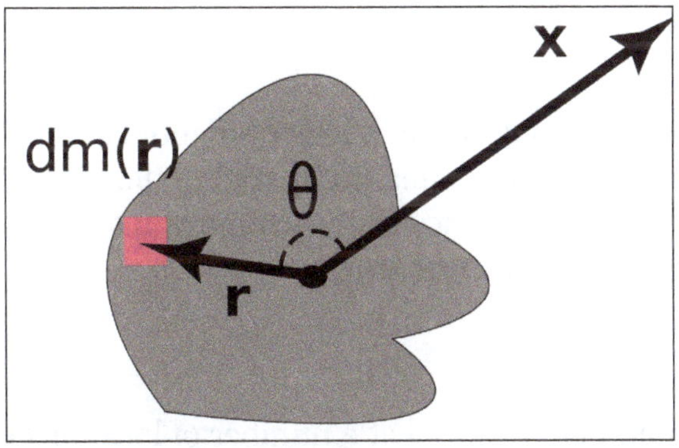

Illustration of a mass distribution (grey) with center of mass as the origin of vectors **x** and **r** and the point at which the potential is being computed at the tail of vector **x**.

The potential can be expanded in a series of Legendre polynomials. Represent the points **x** and **r** as position vectors relative to the center of mass. The denominator in the integral is expressed as the square root of the square to give,

$$V(\mathbf{x}) = -\int_{\mathbb{R}^3} \frac{G}{\sqrt{|\mathbf{x}|^2 - 2\mathbf{x}\cdot\mathbf{r} + |\mathbf{r}|^2}}\, dm(\mathbf{r})$$

$$= -\frac{1}{|\mathbf{x}|}\int_{\mathbb{R}^3} G \Big/ \sqrt{1 - 2\frac{r}{|\mathbf{x}|}\cos\theta + \left(\frac{r}{|\mathbf{x}|}\right)^2}\, dm(\mathbf{r})$$

where, in the last integral, r = |**r**| and θ is the angle between **x** and **r**.

The integrand can be expanded as a Taylor series in $Z = r/|\mathbf{x}|$, by explicit calculation of the coefficients. A less laborious way of achieving the same result is by using the generalized binomial theorem. The resulting series is the generating function for the Legendre polynomials:

$$\left(1-2XZ+Z^2\right)^{-\frac{1}{2}} = \sum_{n=0}^{\infty} Z^n P_n(X)$$

valid for $|X| \leq 1$ and $|Z| < 1$. The coefficients P_n are the Legendre polynomials of degree n. Therefore, the Taylor coefficients of the integrand are given by the Legendre polynomials in $X = \cos\theta$. So the potential can be expanded in a series that is convergent for positions **x** such that $r < |\mathbf{x}|$ for all mass elements of the system (i.e., outside a sphere, centered at the center of mass, that encloses the system):

$$V(\mathbf{x}) = -\frac{G}{|\mathbf{x}|}\int \sum_{n=0}^{\infty}\left(\frac{r}{|\mathbf{x}|}\right)^n P_n(\cos\theta)dm(\mathbf{r})$$

$$= -\frac{G}{|\mathbf{x}|}\int\left(1+\left(\frac{r}{|\mathbf{x}|}\right)\cos\theta+\left(\frac{r}{|\mathbf{x}|}\right)^2\frac{3\cos^2\theta-1}{2}+\cdots\right)dm(\mathbf{r})$$

The integral $\int r\cos\theta dm$ is the component of the center of mass in the **x** direction; this vanishes because the vector **x** emanates from the center of mass. So, bringing the integral under the sign of the summation gives,

$$V(\mathbf{x}) = -\frac{GM}{|\mathbf{x}|} - \frac{G}{|\mathbf{x}|}\int\left(\frac{r}{|\mathbf{x}|}\right)^2\frac{3\cos^2\theta-1}{2}dm(\mathbf{r})+\cdots$$

This shows that elongation of the body causes a lower potential in the direction of elongation, and a higher potential in perpendicular directions, compared to the potential due to a spherical mass, if we compare cases with the same distance to the center of mass. (If we compare cases with the same distance to the *surface*, the opposite is true.)

Numerical Values

The absolute value of gravitational potential at a number of locations with regards to the gravitation from the Earth, the Sun, and the Milky Way is given in the following table; i.e. an object at Earth's surface would need 60 MJ/kg to "leave" Earth's gravity field, another 900 MJ/kg to also leave the Sun's gravity field and more than 130 GJ/kg to leave the gravity field of the Milky Way. The potential is half the square of the escape velocity.

Location	W.r.t. Earth	W.r.t. Sun	W.r.t. Milky Way
Earth's surface	60 MJ/kg	900 MJ/kg	≥ 130 GJ/kg
LEO	57 MJ/kg	900 MJ/kg	≥ 130 GJ/kg
Voyager 1 (17,000 million km from Earth)	23 J/kg	8 MJ/kg	≥ 130 GJ/kg
0.1 light-year from Earth	0.4 J/kg	140 kJ/kg	≥ 130 GJ/kg

Compare the gravity at these locations.

Kepler's Laws of Planetary Motion

Kepler's Law states that the planets move around the sun in elliptical orbits with the sun at one focus. There are three different Kepler's Laws.

Kepler's three Law:

1. Kepler's Law of Orbits: The Planets move around the sun in elliptical orbits with the sun at one of the focii.
2. Kepler's Law of Areas: The line joining a planet to the Sun sweeps out equal areas in equal interval of time.
3. Kepler's Law of Periods: The square of the time period of the planet is directly proportional to the cube of the semimajor axis of its orbit.

1st Law of Orbits

This law is popularly known as the law of orbits. The orbit of any planet is an ellipse around the Sun with Sun at one of the two foci of an ellipse. We know that planets revolve around the Sun in a circular orbit. But according to Kepler, he said that it is true that planets revolve around the Sun, but not in a circular orbit but it revolves around an ellipse. In an ellipse, we have two focus. Sun is located at one of the foci of the ellipse.

2nd Law of Areas

This law is known as the law of areas. The line joining a planet to the Sun sweeps out equal areas in equal interval of time. The rate of change of area with time will be constant. We can see in the above figure, the Sun is located at the focus and the planets revolve around the Sun.

Assume that the planet starts revolving from point P_1 and travels to P_2 in a clockwise direction. So it revolves from point P_1 to P_2, as it moves the area swept from P_1 to P_2 is Δt. Now the planet moves future from P3 to P4 and the area covered is Δt.

As the area traveled by the planet from P_1 to P_2 and P_3 to P_4 is equal, therefore this law is known as the Law of Area. That is the aerial velocity of the planets remains constant. When a planet is nearer to the Sun it moves fastest as compared to the planet far away from the Sun.

3rd Law of Periods

This law is known as the law of Periods. The square of the time period of the planet is directly proportional to the cube of the semimajor axis of its orbit.

$$T^2 \propto a^3$$

That means the time 'T' is directly proportional to the cube of the semi major axis i.e. 'a'. Let us derive the equation of Kepler's 3rd law. Let us suppose,

- m = mass of the planet,

- M = mass of the Sun,
- v = velocity in the orbit.

So, there has to be a force of gravitation between the Sun and the planet.

$$F = \frac{GmM}{r^2}$$

Since it is moving in an elliptical orbit, there has to be a centripetal force.

$$F_c = \frac{mv^2}{r^2}$$

Now, $F = F_c$

$$\Rightarrow \frac{GM}{r} = v^2$$

$$\text{Also, } v = \frac{circum\,ference}{time} = \frac{2\pi r}{t}$$

Combining the above equations, we get

$$\Rightarrow \frac{GM}{r} = \frac{4\pi^2 r^2}{T^2}$$

$$T^2 = \frac{4\pi^2 r^3)}{GM}$$

$$\Rightarrow T^2 \propto r^3$$

References

- Gravity-physics, science: britannica.com, Retrieved 30 April, 2019
- Grøn, øyvind; hervik, sigbjorn (2007). Einstein's general theory of relativity: with modern applications in cosmology (illustrated ed.). Springer science & business media. P. 180. Isbn 978-0-387-69200-5
- 2018 codata value: newtonian constant of gravitation". The nist reference on constants, units, and uncertainty. Nist. 20 may 2019. Retrieved 20 may 2019
- Halliday, david; resnick, robert; walker, jearl (september 2007). Fundamentals of physics (8th ed.). P. 336. Isbn 978-0-470-04618-0
- Fixler, j. B.; foster, g. T.; mcguirk, j. M.; kasevich, m. A. (5 january 2007). "atom interferometer measurement of the newtonian constant of gravity". Science. 315 (5808): 74–77. Bibcode:2007sci...315...74f. Doi:10.1126/science.1135459. Pmid 17204644
- Thornton, stephen t.; marion, jerry b. (2003), classical dynamics of particles and systems (5th ed.), brooks cole, isbn 978-0-534-40896-1
- Mould, j.; uddin, s. A. (10 april 2014). "constraining a possible variation of g with type ia supernovae". Publications of the astronomical society of australia. 31: e015. Arxiv:1402.1534. Bibcode:2014pasa...31...15m. Doi:10.1017/pasa.2014.9
- Keplers-law, gravitation, physics, guides: toppr.com, Retrieved 14 July, 2019

Permissions

All chapters in this book are published with permission under the Creative Commons Attribution Share Alike License or equivalent. Every chapter published in this book has been scrutinized by our experts. Their significance has been extensively debated. The topics covered herein carry significant information for a comprehensive understanding. They may even be implemented as practical applications or may be referred to as a beginning point for further studies.

We would like to thank the editorial team for lending their expertise to make the book truly unique. They have played a crucial role in the development of this book. Without their invaluable contributions this book wouldn't have been possible. They have made vital efforts to compile up to date information on the varied aspects of this subject to make this book a valuable addition to the collection of many professionals and students.

This book was conceptualized with the vision of imparting up-to-date and integrated information in this field. To ensure the same, a matchless editorial board was set up. Every individual on the board went through rigorous rounds of assessment to prove their worth. After which they invested a large part of their time researching and compiling the most relevant data for our readers.

The editorial board has been involved in producing this book since its inception. They have spent rigorous hours researching and exploring the diverse topics which have resulted in the successful publishing of this book. They have passed on their knowledge of decades through this book. To expedite this challenging task, the publisher supported the team at every step. A small team of assistant editors was also appointed to further simplify the editing procedure and attain best results for the readers.

Apart from the editorial board, the designing team has also invested a significant amount of their time in understanding the subject and creating the most relevant covers. They scrutinized every image to scout for the most suitable representation of the subject and create an appropriate cover for the book.

The publishing team has been an ardent support to the editorial, designing and production team. Their endless efforts to recruit the best for this project, has resulted in the accomplishment of this book. They are a veteran in the field of academics and their pool of knowledge is as vast as their experience in printing. Their expertise and guidance has proved useful at every step. Their uncompromising quality standards have made this book an exceptional effort. Their encouragement from time to time has been an inspiration for everyone.

The publisher and the editorial board hope that this book will prove to be a valuable piece of knowledge for students, practitioners and scholars across the globe.

Index

www.ingramcontent.com/pod-product-compliance
Lightning Source LLC
Chambersburg PA
CBHW082042190326
41458CB00010B/3439